Reforestation
21st Century Homestead

Contents

Chapter 1

Reforestation

Tropical tree nursery at Planeta Verde Reforestación S.A.'s planta-tion in Vichada Department, Colombia

Reforestation is the natural or intentional restocking of existing forests and woodlands that have been depleted, usually through deforestation.[*][1] Reforestation can be used to improve the quality of human life by soaking up pollution and dust from the air, rebuild natural habitats and ecosystems, mitigate global warming since forests facilitate biosequestration of atmospheric carbon dioxide, and harvest for resources, particularly timber.

The term *reforestation* is similar to afforestation, the process of restoring and recreating areas of woodlands or forests that may have existed long ago but were deforested or otherwise removed at some point in the past. Sometimes the term *re-afforestation* is used to distinguish between the original forest cover and the later re-growth of forest to an area. Special tools, e.g. tree planting bar, are used to make planting of trees easier and faster.

1.1 Management

Reforestation of large areas can be done through the use of measuring rope (for accurate plant spacing) and dibbers, (or wheeled augers for planting the larger trees) for mak-

A 15-year-old reforested plot of land

A 21-year-old plantation of red pine in Southern Ontario

ing the hole in which a seedling or plant can be inserted. Indigenous soil inoculants (e.g., Laccaria bicolor) can optionally be used to increase survival rates in hardy environments.*[2]

A debatable issue in managed reforestation is whether or not the succeeding forest will have the same biodiversity as the original forest. If the forest is replaced with only one species of tree and all other vegetation is prevented from growing back, a monoculture forest similar to agricultural crops would be the result. However, most reforestation involves the planting of different feedlots of seedlings taken from the area often of multiple species.*[2] Another important factor is the natural regeneration of a wide variety of plant and animal species that can occur on a clear cut. In some areas the suppression of forest fires for hundreds of years has resulted in large single aged and single species forest stands. The logging of small clear cuts and or prescribed burning, actually increases the biodiversity in these areas by creating a greater variety of tree stand ages and species.

1.2   For harvesting

Reforestation need not be only used for recovery of accidentally destroyed forests. In some countries, such as Finland, the forests are *managed* by the wood products and pulp and paper industry. In such an arrangement, like other crops, trees are replanted wherever they are cut. In such circumstances, the industry can cut the trees in a way to allow easier reforestation. The wood products industry systematically replaces many of the trees it cuts, employing large numbers of summer workers for tree planting work. For example, in 2010, Weyerhaeuser reported planting 50 million seedlings.*[3]

In just 20 years, a teak plantation in Costa Rica can produce up to about 400 m of wood per hectare. As the natural teak forests of Asia become more scarce or difficult to obtain, the prices commanded by plantation-grown teak grow higher every year. Other species such as mahogany grow slower than teak in Tropical America but are also extremely valuable. Faster growers include pine, eucalyptus, and *Gmelina*.*[4]

Reforestation, if several native species are used, can provide other benefits in addition to financial returns, including restoration of the soil, rejuvenation of local flora and fauna, and the capturing and sequestering of 38 tons of carbon dioxide per hectare per year.*[5]

The reestablishment of forests is not just simple tree planting. Forests are made up of a diversity of species and they build dead organic matter into soils over time. A major tree-planting program in a place like this would enhance

the local climate and reduce the demands of burning large amounts of fossil fuels for cooling in the summer.*[6]

1.3   For climate change mitigation

Forests are an important part of the global carbon cycle because trees and plants absorb carbon dioxide through photosynthesis. By removing this greenhouse gas from the air, forests function as terrestrial carbon sinks, meaning they store large amounts of carbon. At any time, forests account for as much as double the amount of carbon in the atmosphere.*[7]*:1456 Even as more anthropogenic carbon is produced, forests remove around three billion tons of anthropogenic carbon every year. This amounts to about 30% of all carbon dioxide emissions from fossil fuels. Therefore, an increase in the overall forest cover around the world would tend to mitigate global warming.

There are four major strategies available to mitigate carbon emissions through forestry activities: increase the amount of forested land through a reforestation process; increase the carbon density of existing forests at a stand and landscape scale; expand the use of forest products that will sustainably replace fossil-fuel emissions; and reduce carbon emissions that are caused from deforestation and degradation.*[7]*:1456

Achieving the first strategy would require enormous and wide-ranging efforts. However, there are many organizations around the world that encourage tree-planting as a way to offset carbon emissions for the express purpose of fighting climate change. For example, in China, the Jane Goodall Institute, through their Shanghai Roots & Shoots division, launched the Million Tree Project in Kulun Qi, Inner Mongolia to plant one million trees to stop desertification and help curb climate change.*[8]*[9] China has used 24 billion metres squared of new forest plantation and natural forest regrowth to offset 21% of Chinese fossil fuel emissions in 2000.*[7]*:1456 In Java, Indonesia each newlywed couple is to give whoever is sermonizing their wedding 5 seedlings to combat global warming. Each couple that wishes to have a divorce has to give 25 seedlings to whoever divorces them.*[10]

The second strategy has to do with selecting species for tree-planting. In theory, planting any kind of tree to produce more forest cover would absorb more carbon dioxide from the atmosphere. On the other hand, a genetically modified tree specimen might grow much faster than any other regular tree.*[11]*:93 Some of these trees are already being developed in the lumber and biofuel industries. These fast-growing trees would not only be planted for those industries but they can also be planted to help absorb carbon dioxide faster than slow-growing trees.*[11]*:93

Extensive forest resources placed anywhere in the world will not always have the same impact. For example, large reforestation programs in boreal or subarctic regions have a limited impact on climate mitigation. This is because it substitutes a bright snow-dominated region that reflects the sunlight with dark forest canopies. A study from the National Center for Atmospheric Research in Boulder, Colorado, USA, found that trees in temperate latitudes have a net warming effect on the atmosphere. The heat that dark leaves release without absorbing outweighs the carbon they sequester.*[12] On the other hand, a positive example would be reforestation projects in tropical regions, which would lead to a positive biophysical change such as the formation of clouds. These clouds would then reflect the sunlight, creating a positive impact on climate mitigation.*[7]*:1457

There is an advantage to planting trees in tropical climates with wet seasons. In such a setting, trees have a quicker growth rate because they can grow year-round. Trees in tropical climates have, on average, larger, brighter, and more abundant leaves than non-tropical climates. A study of the girth of 70,000 trees across Africa has shown that tropical forests are soaking up more carbon dioxide pollution than previously realized. The research suggests almost one fifth of fossil fuel emissions are absorbed by forests across Africa, Amazonia and Asia. Simon Lewis, a climate expert at the University of Leeds, who led the study, said: "Tropical forest trees are absorbing about 18% of the carbon dioxide added to the atmosphere each year from burning fossil fuels, substantially buffering the rate of change." *[13]

It is also important to deal with the rate of deforestation. At this point, there are 13 billion metres squared of tropical regions that are deforested every year. There is potential for these regions to reduce rates of deforestation by 50% by 2050, which would be a huge contribution to stabilize the global climate.*[7]*:1456

1.4 Incentives

Some incentives for reforestation can be as simple as a financial compensation. Streck and Scholz (2006) explain how a group of scientists from various institutions have developed a compensated reduction of deforestation approach which would reward developing countries that disrupt any further act of deforestation. Countries that participate and take the option to reduce their emissions from deforestation during a committed period of time would receive financial compensation for the carbon dioxide emissions that they avoided.*[14]*:875 To raise the payments, the host country would issue government bonds or negotiate some kind of loan with a financial institution that would want to take

part in the compensation promised to the other country. The funds received by the country could be invested to help find alternatives to the extensive cutdown of forests. This whole process of cutting emissions would be voluntary, but once the country has agreed to lower their emissions they would be obligated to reduce their emissions. However, if a country was not able to meet their obligation, their target would get added to their next commitment period. The authors of these proposals see this as a solely government-to-government agreement; private entities would not participate in the compensation trades.*[14]*:876

1.5 Examples

Forest regrowth in Mount Baker-Snoqualmie National Forest, Washington state, USA

1.5.1 Africa

Plans to plant a nine mile width of trees on the Southern Border of the Sahara Desert The Great Green Wall initiative is a pan-African proposal to "green" the continent from west to east in order to battle desertification. It aims at tackling poverty and the degradation of soils in the Sahel-Saharan region, focusing on a strip of land of 15 km (9 mi) wide and 7,500 km (4,750 mi) long from Dakar to Djibouti.*[15]

1.5.2 Canada

In Canada, overall forest cover is increasing over the last decades.

1.5.3 China

In China, extensive replanting programs have existed since the 1970s. Programs have had overall success. The forest cover has increased from 12% of China's land area to 16%. However, specific programs have had limited success. The "Green Wall of China", an attempt to limit the expansion of the Gobi Desert is planned to be 2,800 miles (4,500 km) long and to be completed in 2050. China plans to plant 26 billion trees in the next decade that is two trees for every Chinese citizen per year.[16] China requires that students older than 11 years old plant one tree a year until their high school graduation.[17]

1.5.4 Germany

Reforestation is required as part of the federal forest law. 31% of Germany is forested, according to the second forest inventory of 2001–2003. The size of the forest area in Germany increased between the first and the second forest inventory due to forestation of degenerated bogs and agricultural areas.[18]

1.5.5 Lebanon

For thousands of years Lebanon was covered by forests, one particular species of interest, Cedrus libani or **Lebanon Cedar** was exceptionally valuable timber species. Virtually every ancient culture that shared the Mediterranean Sea harvested these trees, the Ancient Egyptians, the Greeks, the Roman Empire, early Christians, the Muslims, Persians, Assyrians and Babylonians. Extensive deforestation has occurred, with only small remnants of the original forests surviving. Deforestation has been particularly severe in Lebanon and on Cyprus. The Lebanon Reforestation Initiative aims to restore Lebanon's native forests. Project's financed locally and by international charity are performing extensive reforestation of cedar being carried out in the Mediterranean region, particularly in Lebanon and Turkey, where over 50 million young cedars are being planted annually.

1.5.6 United States

It is the stated goal of the US Forest Service to manage forest resources sustainably. This includes reforestation after timber harvest, among other programs.[19]

1.5.7 Organizations

Trees for the Future has assisted more than 170,000 families, in 6,800 villages of Asia, Africa and the Americas, to plant over 35 million trees.[20]

Wangari Maathai, 2004 Nobel Peace Prize recipient, founded the Green Belt Movement which planted over 47 million trees to restore the Kenyan environment.[21]

Shanghai Roots & Shoots, a division of the Jane Goodall Institute, launched The Million Tree Project in Kulun Qi, Inner Mongolia to plant one million trees to stop desertification and alleviate global warming.[22][23]

1.5.8 Individual Efforts

1.6 Criticisms

Reforestation competes with other land uses such as food production, livestock grazing and living space for further economic growth. Reforestation often has the tendency to create large fuel loads, resulting in significantly hotter combustion than fires involving low brush or grasses. Reforestation can divert large amounts of water from other activities. Reforesting sometimes results in extensive canopy creation that prevents growth of diverse vegetation in the shadowed areas and generating soil conditions that hamper other types of vegetation. Trees used in some reforesting efforts (e.g., eucalyptus globulus) tend to extract large amounts of moisture from the soil, preventing the growth of other plants.

There is also the risk that through a forest fire or insect outbreak much of the stored carbon in a reforested area could make its way back to the atmosphere.[7]:1456 Reduced harvesting rates and fire suppression have caused an increase in the forest biomass in the western United States over the past century. This causes an increase of about a factor of four in the frequency of fires due to longer and hotter dry seasons.[7]:1456

1.7 See also

- 10,000 Trees for the Rouge Valley, a reforestation program in Toronto, Canada

- Aerial reforestation

- Afforestation

- Deforestation

- Forest gardening

- Forest restoration

- Forestry
- Greenland Arboretum
- Hoedads Reforestation Cooperative
- Jewish National Fund
- Land rehabilitation
- Natural landscape
- Plant A Tree Today Foundation
- Pottiputki (tool)
- Restoration ecology
- Revegetation
- Rewilding (conservation biology)
- Richard St. Barbe Baker
- "The Man Who Planted Trees" (French title *L'homme qui plantait des arbres*), a tale by French author Jean Giono, published in 1953, and made into an animated film in 1987
- Tree credits
- Tree planting
- Tubestock
- Urban reforestation
- Deforestation and climate change

1.8 References

[1] "Reforestation - Definitions from Dictionary.com". dictionary.reference.com. Retrieved 2008-04-27.

[2] http://kennethmarendefoundation.com/index.php/reforestation-problem

[3] "Sustainable Forest Management". *Key Timberland Statistics*. Weyerhaeuser. 10 June 2011. Retrieved 7 January 2012.

[4] "Forest plantation yields in the tropical and subtropical zone". *Forestry Department*. Retrieved 15 February 2014.

[5] "Reforestation and Afforestation". *Green Collar Association*. Retrieved 15 February 2014.

[6] Wood well, G.M.; North, WJ (1988-12-16). "CO_2 Reduction and Reforestation". *Science* (ALAS) **242** (4885): 1493–1494. doi:10.1126/science.242.4885.1493-a. PMID 17788407.

[7] Canadell, J.G.; M.R. Raupach (2008-06-13). "Managing Forests for Climate Change". *Science* (AAAS) **320** (5882): 1456–1457. doi:10.1126/science.1155458. PMID 18556550.

[8] "Shanghai Roots & Shoots -". *jgi-shanghai.org*.

[9] "Million Tree Project :: Home". *mtpchina.org*.

[10] "Five tree fee for a Java wedding". *BBC News*. 2007-12-03. Retrieved 2010-08-29.

[11] "A changing climate of opinion?". *The Economist* (The Economist Newspaper Limited) **387**: 93–96. 2008. Retrieved 2010-08-29.

[12] B.W., Time Magazine, 2007

[13] Adam, David (2009-02-18). "Fifth of world carbon emissions soaked up by extra forest growth, scientists find". *The Guardian* (London). Retrieved 2010-05-22.

[14] Streck, C.; S.M. Scholz (2006). "The role of forests in global climate change: whence we come and where we go". *International Affairs* (The Royal Institute of International Affairs) **82** (5): 861–879. doi:10.1111/j.1468-2346.2006.00575.x.

[15] "Real happenings and Facts". *naturne.blogspot.com*.

[16] http://en.people.cn/90882/7675458.html

[17] ?

[18] "Grußwort - Bundeswaldinventur". *bundeswaldinventur.de*.

[19] "Forest Service Chief testifies before Senate appropriations committee on 2013 agency budget". US Forest Service. 18 April 2012. Retrieved 29 April 2012.

[20] Trees for the Future

[21] "The Greenbelt Movement".

[22] "Shanghai Roots & Shoots".

[23] "The Million Tree Project".

1.9 Further reading

- Bonan, G. B. (2008). "Forests and climate change: Forcings, feedbacks, and the climate benefits of forests". *Science* **320** (5882): 1444–1449. doi:10.1126/science.1155121. PMID 18556546.

- Scheil, D.; Murdiyarso, D. (2009). "How Forests Attract Rain: An Examination of a New Hypothesis". *BioScience* **59** (4): 341–347. doi:10.1525/bio.2009.59.4.12.

1.10 External links

- Tropical Reforestation

- Plant a Tree and Become Carbon Neutral

- Brazilian "Mata Atlantica" reforestation initiative

- Reforestation Carbon Offsets

- "Perpetual Timber Supply Through Reforestation as Basis For Industrial Permanency: The Story Of Bogalusa" By Courtenay De Kalb, July 1921

- Saimiri Wildlife; Reforestation for endangered wildlife, side provides many pictures.

- A tree a day, keeps the carbon away

- Trees and climate change: a practical guide for woodland owners and managers

- Floresta, a Christian nonprofit with a reforestation and poverty world mission.

- Plant with Purpose

- Shanghai Roots & Shoots - Million Tree Project

- Reforestation Information

Chapter 2

Urban reforestation

Urban reforestation is the practice of planting trees, typically on a large scale, in urban environments.[1] It sometimes includes also urban horticulture and urban farming.[2] Reasons for practicing urban reforestation include urban beautification,[1] increasing shade,[1] modifying the urban climate,[3] improving air quality,[4] and restoration of urban forests after a natural disaster.[5]

2.1 Programs

Large scale urban reforestation programs include New York City's Million Tree Initiative,[6] and TreePeople in Los Angeles, which planted 1 million trees in preparation for the 1984 Summer Olympics and continued planting thereafter.[1]

Grassroots efforts include Friends of the Urban Forest in San Francisco which advocates the planting of street trees[1] and the *Urban Reforestation* organization in Australia, which focuses on sustainable living in urban places.[2]

2.2 Criticisms

Urban reforestation efforts compete for money and urban land that could be used for other purposes. For example, effort placed in planting new trees can take away from maintenance of already established trees.[6]

2.3 See also

- Reforestation
- Tree planting
- Urban agriculture
- Urban forest
- Urban forestry
- Urban heat island

2.4 References

[1] Gary Moll, Sara Ebenreck (1989). *Shading Our Cities: A Resource Guide For Urban And Community Forests*. Island Press. ISBN 9780933280953.

[2] Green thumbs and high-rise ambitions, *The Age*, June 11, 2010, See also urbanreforestation.com website.

[3] Hall, Justine M.; John F. Handley; A. Roland Ennos (15 March 2012). "The potential of tree planting to climate-proof high density residential areas in Manchester, UK". *Landscape and Urban Planning* **104** (3-4): 410–417. doi:10.1016/j.landurbplan.2011.11.015.

[4] Taha, Halder (May 2008). "Urban Surface Modification as a Potential Ozone Air-quality Improvement Strategy in California: A Mesoscale Modelling Study". *Boundary-Layer Meteorology* **127** (2): 219–239. doi:10.1007/s10546-007-9259-5.

[5] Lisa L. Burban, John W. Anderson (1996). *Storms Over the Urban Forest: Planning, Responding, and Regreening - A Community Guide to Natural Disaster Relief*. DIANE Publishing. ISBN 9780788129483.

[6] Corso, Phil. "Avella opposes mayor's Million Trees effort". TimesLedger. Retrieved 31 January 2013.

Chapter 3

Afforestation

An afforestation project in Rand Wood, Lincolnshire, England

Afforestation is the establishment of a forest or stand of trees in an area where there was no forest.[*][1] Reforestation is the reestablishment of forest cover, either naturally (by natural seeding, coppice, or root suckers) or artificially (by direct seeding or planting).[*][2] Many governments and non-governmental organizations directly engage in programs of *afforestation* to create forests, increase carbon capture and sequestration, and help to anthropogenically improve biodiversity. (In the UK, afforestation may mean converting the legal status of some land to "royal forest".) Special tools, e.g. tree planting bar, are used to make planting of trees easier and faster.

3.1 Biological process

Main article: Gap dynamics

Gap dynamics refers to the pattern of plant growth that occurs following the creation of a forest gap, a local area of natural disturbance that results in an opening in the canopy of a forest. Gap dynamics are a typical characteristic of both temperate and tropical forests and have a wide variety of causes and effects on forest life.

3.2 In areas of degraded soil

In some places, forests need help to reestablish themselves because of environmental factors. For example, in arid zones, once forest cover is destroyed, the land may dry and become inhospitable to new tree growth. Other factors include overgrazing by livestock, especially animals such as goats, cows, and over-harvesting of forest resources. Together these may lead to desertification and the loss of topsoil; without soil, forests cannot grow until the long process of soil creation has been completed - if erosion allows this. In some tropical areas, forest cover removal may result in a duricrust or duripan that effectively seal off the soil to water penetration and root growth. In many areas, reforestation is impossible because people are using the land. In other areas, mechanical breaking up of duripans or duricrusts is necessary, careful and continued watering may be essential, and special protection, such as fencing, may be needed.

3.3 Countries and regions

Afforested botanical garden in Hattori Ryokuchi Park, Japan.

3.3.1 Brazil

Due to the extensive and ongoing Amazon deforestation of the last few decades,[*][3] the small efforts of afforestation are insignificant on a national scale of the Amazon Rainforest.[*][4]

3.3.2 China

China has deforested most of its historically wooded areas. China reached the point where timber yields declined far below historic levels, due to over-harvesting of trees beyond sustainable yield.[*][5] Although it has set official goals for reforestation, these goals are set over an 80-year time horizon and have not been significantly met by 2008. China is trying to correct these problems by projects as the Green Wall of China, which aims to replant a great deal of forests and halt the expansion of the Gobi desert. A law promulgated in 1981 requires that every school student over the age of 11 plants at least one tree per year. As a result, China currently has the highest afforestation rate of any country or region in the world, with 47,000 square kilometers of afforestation in 2008.[*][6] However, the forest area per capita is still far lower than the international average.[*][7] There has also been considerable criticism regarding the effectiveness of planting so many trees especially in regions where they never grew prior. Studies reveal that the water table of those areas is becoming deeper indicating significant water loss.

3.3.3 India

Afforestation in South India

India has witnessed a minor increase in the percentage of the land area under forest cover from 1950 to 2006. In 1950 around 40.48 million hectares was covered by forest. In 1980 it increased to 67.47 million hectares and in 2006 it was found to be 69 million hectares. 23% of India is covered by forest.[*][8] The forests of India are grouped into 5 major categories and 16 types based on biophysical criteria. 38% of forest is categorised as subtropical dry deciduous and 30% as tropical moist deciduous plus other smaller groups. It is taken care that only local species are planted in an area. Trees bearing fruits are preferred wherever due to their function as a food source.

3.3.4 Hong Kong

Since the founding of the crown colony in the 19th century, afforestation has taken place to prevent soil erosion in the catchment areas of the reservoirs that were built. During the Japanese occupation in the Second World War, the countryside was deforested as the remaining population required fuel to survive. Most of the trees were cut down and extensive reafforestation was carried out after the war. Trees that were planted are mostly non-native species, such as: Pinus massoniana, Acacia confusa (Formosan acacia), Lophostemon confertus and the Paper Bark Tree.

3.3.5 Burkina-Faso

Desertification is increasing along the Sahel, the strip of land between Africa's fertile tropics and the Sahara Desert. After a crippling famine in the 1970s caused by overgrazing and deforestation, a local community approach has been pioneered by Yacouba Sawadogo, a peasant farmer.[*][9] By replanting trees and crops together in holes filled with compost, whole villages have been able to move back to areas considered uninhabitable.

3.3.6 Iran

Iran is considered a low forest cover region of the world with present cover approximating seven percent of the land area. This is a value reduced by an estimated six million hectares of virgin forest, which includes oak, almond and pistachio.[*][10] Due to soil substrates, it is difficult to achieve afforestation on a large scale compared to other temperate areas endowed with more fertile and less rocky and arid soil conditions.[*][10] Consequently, most of the afforestation is conducted with non-native species,[*][10] leading to habitat destruction for native flora and fauna, and resulting in an accelerated loss of biodiversity.[*][3]

JNF trees in the Negev Desert. Man-made dunes (here a liman) help keep in rainwater, creating an oasis.

3.3.7   Japan

Main article: Afforestation in Japan

3.3.8   Israel

Main article: Jewish National Fund § Afforestation

Tree-planting is an ancient Jewish tradition, mentioned in the Talmud as being more important than greeting the Messiah.[11] With over 240 million planted trees, Israel is one of only two countries that entered the 21st century with a net gain in the number of trees, due to massive afforestation efforts.[12] Israeli forests are the product of a major afforestation campaign by the Jewish National Fund (JNF).[13]

Critics argue that many JNF lands inside the West Bank were illegally confiscated from Palestinian refugees, and that the JNF furthermore should not be involved with lands in the West Bank.[14] Shaul Ephraim Cohen has claimed that trees have been planted to restrict Bedouin herding.[15] Susan Nathan wrote that forests were planted on the site of abandoned Arab villages after the 1948 war.[16]

Since 2009, the JNF has provided the Palestinian Authority with 3,000 tree seedlings for a forested area being developed on the edge of the new city of Rawabi, north of Ramallah.[17]

3.3.9   North Africa

In North Africa, the Sahara Forest Project coupled with the Seawater greenhouse has been proposed. Some projects have also been launched in countries as Senegal to revert

desertification. As of 2010, African leaders are discussing the combining of national resources to increase effectiveness.[18] In addition, other projects as the Keita Project in Niger have been launched in the past, and have been able to locally revert damage done by desertification. See Development aid#Effectiveness

3.3.10   Europe

Europe has deforested the majority of its historical forests. The European Union (EU) has paid farmers for afforestation since 1990, offering grants to turn farmland back into forest and payments for the management of forest. Between 1993 and 1997, EU afforestation policies made possible the re-forestation of over 5,000 square kilometres of land. A second program, running between 2000 and 2006, afforested more than 1000 square kilometres of land (precise statistics not yet available). A third such program began in 2007. Europe's forests are growing by 0.8 million ha a year thanks to these programmes.[19]

In Poland, the National Program of Afforestation was introduced by the government after World War II, when area of forests shrank to 20% of country's territory. Consequently, forested areas of Poland grew year by year, and on December 31, 2006, forests covered 29% of the country (see: Polish forests). It is planned that by 2050, forests will cover 33% of Poland.

According to FAO statistics, Spain had the third fastest afforestation rate in Europe in the 1990-2005 period, after Iceland and Ireland.[20][21] In those years, a total of 44,360 square kilometers were afforested, and the total forest cover rose from 13,5 to 17,9 million hectares. In 1990, forests covered 26,6% of the Spanish territory. As of 2007, that figure had risen to 36,6%. Spain today has the fifth largest forest area in the European Union.[22]

In January 2013 the UK government set a target of 12% woodland cover in England by 2060, up from the then 10%.[23] Government-backed initiatives such as the Woodland Carbon Code are intended to support this objective by encouraging corporations and landowners to create new woodland to offset their carbon omissions.

3.3.11   Australia

In Adelaide, South Australia (a city of 1.3 million) Premier Mike Rann (2002 to 2011) launched an urban forest initiative in 2003 to plant 3 million native trees and shrubs by 2014 on 300 project sites across the metro area. The projects range from large habitat restoration projects to local biodiversity projects. Thousands of Adelaide citizens have participated in community planting days. Sites include parks, reserves, transport corridors, schools, water courses and coastline Only trees native to the local area are planted to ensure genetic integrity. Premier Rann said the project aimed to beautify and cool the city and make it more live-able; improve air and water quality and reduce Adelaide's greenhouse gas emissions by 600,000 tonnes of C02 a year. He said it was also about creating and conserving habitat for wildlife and preventing species loss.*[24]

3.3.12   United States

The United States is roughly one-third covered in forest and woodland. Nevertheless, areas in the US were subject to significant tree planting. In the 1800s people moving west-ward encountered the Great Plains; land with fertile soil, a growing population and a demand for timber but with few trees to supply it. So tree planting was encouraged along homesteads. Arbor Day was founded in 1872 by Julius Ster-ling Morton in Nebraska City, Nebraska. By the 1930s the environmental disaster, the Dust Bowl signified a reason for significant tree cover. Public work's programs under the New Deal saw the planting of 18,000 miles of windbreaks stretching from North Dakota to Texas to fight soil erosion (see Great Plains Shelterbelt).

Billion Trees Initiative

At their summit in Copenhagen in 2009, organised by the UK based The Climate Group, leaders of sub-national gov-ernments - States, Regions and Provinces - unanimously supported a recommendation by Premier Rann to plant 1 billion trees across their varied jurisdictions. The initia-tive was strongly supported by leaders present including Quebec Premier Jean Charest, California Governor Arnold Schwarzenegger and Scottish First Minister Alex Salmond. At a subsequent meeting in Rio de Janeiro in June 2012, The Climate Group announced that it had already received commitments by member governments to plant more than 500 million trees.*[25]

3.4   See also

- Agroforestry
- Buffer strip
- CarbonFix Standard
- Deforestation
- Desertification
- Forestry
- Great Plains Shelterbelt
- Groasis Waterboxx
- International Year of Forests
- Japanese afforestation
- Reforestation
- Sand fence
- Seawater greenhouse
- Tubestock
- Windbreak
- Deforestation and climate change

3.5   References

3.5.1   Notes

[1]   "SAFnet Dictionary | Definition For [afforestation]". Dictionaryofforestry.org. 2008-10-23. Retrieved 2012-02-17.

[2]   "SAFnet Dictionary | Definition For [reforestation]". Dictionaryofforestry.org. 2008-08-13. Retrieved 2012-02-17.

[3]   E. O. Wilson, 2002

[4]   A.Cattaneo, 2002

[5]   G.A.McBeath, 2006

[6]   "China to plant more trees in 2009_English_Xinhua". News.xinhuanet.com. 2009-01-09. Retrieved 2012-02-17.

[7]   "51.54 billion trees planted by ordinary Chinese in 27 years - People's Daily Online". English.people.com.cn. 2008-03-11. Retrieved 2012-02-17.

[8]   "India: Environmental Profile". *rainforests.mongabay.com*. Retrieved 2015-05-04.

[9]   Wildash, Paula. "1080 Films - The Man Who Stopped The Desert". *www.1080films.co.uk*. Retrieved 2015-05-04.

[10]   J.A.Stanturf, 2004

[11]   "President of German States Council of Education Minis-ters Plants Tree at Kennedy Memorial". JPost.com: Green Israel: People And The Environment. *Jerusalem Post* (The Jerusalem Post). July 29, 2009. Retrieved December 13, 2013.

[12]   "Israel Forestry & Ecology". Jewish National Fund, East 69th Street, NY 10021 USA. Retrieved 29 October 2011.

[13]   "JNF Tree Planting Center". Jewish National Fund, East 69th Street, NY 10021, USA. Retrieved 29 October 2011.

[14] Dan Leon."The Jewish National Fund: How the Land Was 'Redeemed': The JNF's historical concept of exclusively Jewish land is wholly anachronistic"; Palestine-Israel Journal, Vol 12 No. 4 & Vol 13 No. 1, 05/06

[15] Shaul Ephraim Cohen. "The Politics of Planting"; University of Chicago 1993 p.121

[16] Nathan, Susan (2005). The Other Side of Israel: My Journey Across the Jewish/Arab Divide. New York: Nan A. Talese. pp. 130–131. ISBN 978-0-385-51456-9.

[17] Gross, Tom (2009-12-02). "Building Peace Without Obama's Interference". Online.wsj.com. Retrieved 2013-12-21.

[18] "Combining of green walls". Afriqueavenir.org. Archived from the original on 2010-07-18. Retrieved 2012-08-26.

[19] "European Wood_Forest growth". *www.europeanwood.org.cn*. Retrieved 2015-05-04.

[20] "FAO Data". Blatantworld.com. Retrieved 2012-08-26.

[21] "Mongabay.com: Deforestation tables and charts for Spain". Rainforests.mongabay.com. Retrieved 2012-08-26.

[22] "United Nations Statistics Division - Environment Statistics". Unstats.un.org. Retrieved 2012-02-17.

[23] "Government Forestry and Woodlands" (PDF). Defra. Retrieved 13 June 2013.

[24] http://www.milliontrees.com.au

[25] http://www.theclimategroup.org

3.5.2 Bibliography

- Cattaneo, Andrea (2002) *Balancing Agricultural Development and Deforestation in the Brazilian Amazon*, Int Food Policy Res Inst IFPRI, 146 pages ISBN 0-89629-130-8

- Heil, Gerrit W., Bart Muys and Karin Hansen (2007) *Environmental Effects of Afforestation in North-Western Europe*, Springer, 320 pages ISBN 1-4020-4567-0

- Halldorsson G., Oddsdottir, ES and Sigurdsson BD (2008) *AFFORNORD Effects of Afforestation on Ecosystems, Landscape and Rural Development*, TemaNord 2008:562, 120 pages ISBN 978-92-893-1718-4

- Halldorsson G., Oddsdottir, ES and Eggertsson O (2007) *Effects of Afforestation on Ecosystems, Landscape and Rural Development. Proceedings of the AFFORNORD conference, Reykholt, Iceland, June 18–22, 2005*, TemaNord 2007:508, 343pages ISBN 978-92-893-1443-5

- McBeath, Gerald A., and Tse-Kang Leng (2006) *Governance of Biodiversity Conservation in China and Taiwan*, Edward Elgar Publishing, 242 pages ISBN 1-84376-810-0

- Stanturf, John A. and Palle Madsen (2004) *Restoration of Boreal and Temperate Forests*, CRC Press, 569 pages ISBN 1-56670-635-1

- Wilson, E. O. (2002) *The Future of Life*, Vintage ISBN 0-679-76811-4

3.6 External links

- Raunet Michel and Naudin Krishna, 2006. Combating desertification through direct seeding mulch-based cropping systems (DMC). Les dossiers thématiques du CSFD. Issue 4. 40 pp.

Chapter 4

Afforestation in Japan

Afforested botanical garden in Hattori Ryokuchi Park, Japan

The Japanese temperate rainforest is well sustained and maintains a high biodiversity. One method that has been utilized in maintaining the health of forests in **Japan** has been **afforestation**. The Japanese government and private businesses have set up multiple projects to plant native tree species in open areas scattered throughout the country. This practice has resulted in shifts in forest structure and a healthy temperate rainforest that maintains a high biodiversity.

4.1 Purpose of afforestation

The primary goal of afforestation projects in Japan is to develop the forest structure of the nation and to maintain the biodiversity found in the Japanese wilderness. The Japanese temperate rainforest is scattered throughout the Japanese archipelago and is home to many endemic species that are not naturally found anywhere else. As development of the country's caused a decline in forest cover, a reduction in biodiversity was seen in those areas.[1] In an effort to counteract the observed decline in biodiversity, Japan began many afforestation projects. New tree stands were planted all over the archipelago and native species that in-

habited the existing wild forests began to occupy the newly forested areas.

Afforestation projects in Japan first started after the rebuilding that followed World War II. In efforts to restore the country's infrastructure after the war, large areas of forest were clear-cut for timber and to create pastures to attract immigrant farmers.[2] A new management plan for the forests of Japan was instated after many pastures were abandoned and there was a recognized massive decline of old growth and secondary forests.[3] Forest plantations were created to increase the health of Japanese forests and to sustain the nation's timber industry. Afforestation was combined with changes in logging practices that called for reduced clear-cutting and low impact logging over a larger area.[4]

Many private businesses in the country take part in other afforestation projects as a means of reducing the carbon emissions of the company. Carbon sequestration is a major incentive for businesses to plant seedlings and saplings that will store atmospheric carbon in their biomass as they grow. Companies like Japex and Toyota have planted and maintained tree stands beside their plants located across Japan; planting several thousand native trees in an effort to offset carbon emissions. Businesses also monitor the health of newly planted tree stands by tracking growth and surveying near-by wild forests as a comparison of how well the new stands are doing.

Historically, the forests of Japan were not extensively cut as a means of reducing the frequency of landslides and other natural disasters occurring. The root structure of the forests held the soil in place and stabilized the ground in an environment that experiences heavy rainfall and earthquake activity. In present afforestation projects, trees are used to anchor the soil and reduce the amount of soil erosion undergone by that area. A healthy tree stand also restricts the amount of soil disturbance due to rainfall by intercepting heavy precipitation in the canopy and by slowing the rate of run-off with the litter covering the soil. The role that the Japanese temperate rainforest plays in the prevention of soil erosion was recognized early on in the history of the nation and continues to be a positive reason to reforest open areas as new tree stands reduce the amount of nutrients lost in an area and allow for a more productive ecosystem.

4.2 History

The logging industry of Japan can be divided into pre-World War II and post-WWII. Prior to the Second World War, Japan has a small timber industry that impacted a small area as most of the forests were preserved to prevent erosion and maintain the overall quality of the land. It was during World War II that the logging industry really expanded as timber was increasingly in demand.

The increased demand of timber during the war continued on after it had ended. Japan continued to cut down large forested areas in order to rebuild. Old growth and secondary stands were increasingly fragmented as areas were clear-cut to allow for reconstruction.*[5] Between 1945 to 1965, there was rapid deforestation throughout the temperate rainforest of Japan.

During this twenty-year period, open pastures were cultivated in the Hokkaido region of northern Japan in an effort to attract more people to the area for agricultural development. Over forty-five thousand households immigrated to the Hokkaido region, but only 28.6% stayed.*[6] Most of the cultivated pasture land was abandoned and returned to the Japanese government from 1966-1977, as climate conditions in the area were not conducive to good crop yield. The shift from old growth forest to pasture left large areas of reduced soil fertility that trees were unable to recolonize. Due to a lack of a seed bank and competition with dwarf bamboo, human involvement was necessary to reforest the area. From 1978-2005 native trees with high growth rates were planted in plantations. It was mostly conifers that were planted in the area, but it has aided in the recovery of a conifer-broadleaf mixed forest.

In 1973, there was new forestry management implemented in Japan. It restricted clear-cutting and called for selective cutting over a large area. Wild forest was left surrounding areas that had been cut and newly planted.*[7] Though the plantation stage converted most of the old growth and secondary forests to conifer plantations, the new management pushed for reduced that shift.

Present day Japan continues to organize several afforestation programs throughout the nation. The government does continue to regulate forestry, but most of the afforestation projects are now put on by various private organizations. Also, most of the timber that Japan uses is imported from foreign markets; this allows for reduced logging in Japan and more time for Japanese forests to grow.

4.3 Processes involved

There has been varying degrees of human involvement in afforested regions. For most of the areas that are reforested, there is a clearing of land that is planted with native saplings. Growth is generally monitored, but there is not much human involvement beyond checking the health of a stand.

In other regions, there is more land preparation involved prior to planting anything. Most open areas get quickly colonized by dwarf bamboo that out-competes any saplings trying to grow in the area. To give any new saplings a chance to grow in a new area people practice soil scarification. The land is plowed and soil is rotated. This process kills off any tall grasses of bamboo in the area, effectively reducing competition for new saplings. This process does subject the soil to compaction, making it harder for roots to establish; however, once the trees are established, they can regenerate and facilitate growth of other tree species.*[8]

To facilitate further growth and development in plantations, there is selective logging in order to create microhabitats for new growth using pit-and-mound topography. Trees are tipped over to create mounds that turn up more nutrients in the soil and allow for more sunlight in some patches of the understory. A healthy understory of young trees serves as a source of regeneration when an older, larger tree dies.*[9]*[10]

4.4 Tree species

Japan afforestation projects plant only tree species native to Japan. Originally, species were chosen for their rapid growth rates and tolerance for multiple environmental conditions, which led to a large shift from old growth and secondary forests to conifer plantations. Current afforestation projects plant a more diverse number of species.

The dominant tree species planted are Japanese cedar (*Cryptomeria japonica*) and Japanese cypress (*Chamaecyparis obtusa*). Both conifer species are native to Japan and prefer to grow in deep, well-drained soils in warm, wet climates. Both species were able to successfully occupy the open areas made available by rapid deforestation. Other than being able to grow quickly, the trees were chosen because of their ornamental use in urban areas and because their wood is highly valuable timber.*[11]

Other native tree species are planted in an effort to regenerate mixed forests of conifers and broadleaf deciduous trees. Other species include: Mongolian oak (*Quercus mongolica*), lobed elm (*Ulmus laciniata*), Japanese white birch (*Betula platyphylla*), Glehn's spruce (*Picea glehnii*), and sakhalin fir (*Abies sachalinensis*).

4.5 Changes in soil

One study examining the soil of various Japanese conifer plantations found that the plantations had a higher concentration of soil organic carbon than wild, natural forests in Japan.[*][12] Soil carbon is associated with the amount of nutrients that a soil sample can hold. When areas undergo deforestation a lot of the soil organic carbon is exposed and vulnerable to degradation. This results in a loss of nutrients in deforested areas. It would be expected that plantations would have a much smaller amount of organic carbon in the soil because the soil was exposed and experienced a nutrient loss; however, it was found that conifer plantations in Japan actually had more soil carbon than wild forests in the area. It was determined that the heavy litter from newer conifer plantations resulted in an abundance of soil carbon. It was also noted that there was less soil carbon found in plantations that had undergone soil scarification prior to being planted. This was determined to be because soil scarification brought up and exposed more soil organic carbon and more was lost to degradation; however there was still a high concentration of soil carbon in those plantations as well.

4.6 Concerns

Afforestation is generally seen as a positive thing, and the reestablishment of forests in Japan has been positive overall. Large-scale afforestation is still a new concept, and there are some concerns associated with it. Afforested regions have impacts on local watersheds as well as species interactions.

One study examined the reduced water quality of watersheds downstream of afforested conifer plantations.[*][13] The concern was that there was no improvement seen with public water despite improved water treatment in those areas. It was found that heavy rainfall upstream in afforested conifer plantations caused an increase in nitrogen and phosphorus running off into the streams. The undeveloped understory in those plantations means that a lot of the soil is exposed. The litter does slow down the rate of the run-off, but a large amount of nutrients still get picked up and put into the watershed. This resulted in an increase in particulate and dissolved nitrogen and phosphorus.

A large concern in Japan is root rot experienced in conifer species as a result of *Armillaria*, a group of fungus species found in Asia. Sites all over Japan were examined and it was found that every conifer species native to Japan was vulnerable to at least one species of *Armillaria*.[*][14] This is a major concern because every conifer species is vulnerable, but it is especially concerning because the majority of the Japanese temperate rainforest is made up of conifer

plantations consisting of monocultures of just a few species. Another concern is due to Japan's practice of importing timber from other places. Although this has proven to be beneficial for Japanese forests, it has resulted in massive deforestation in places that do not have well-developed regulation on logging. Japan is one of the largest importers of timber, which has come at a cost of temperate and tropical forest decline in China and South America. As a result of this negativity, Japan has organized afforestation programs in other countries.

4.7 See also

- Deforestation and climate change

4.8 References

4.8.1 Notes

[1] Miyamoto et al. 2008

[2] Shoyama 2008

[3] Miyamoto et al. 2008

[4] Miyamoto et al. 2008

[5] Miyamoto et al. 2008

[6] Shoyama 2008

[7] Miyamoto et al. 2008

[8] Resco de Dios et al. 2005

[9] Resco de Dios et al. 2005

[10] Naguchi et al. 2003.

[11] Miyamoto et al. 2008.

[12] Hirotsugu et all. 2010.

[13] Ide et al. 2007.

[14] Hasegawa et al. 2011.

4.8.2 Bibliography

1. Arai, Hirotsugo, and Naoko Tokuchi. "Soil Organic Carbon Accumulation Following Afforestation in a Japanese Coniferous Plantation Based on Particle-Size Fractionation and Stable Isotope Analysis." Geoderma 159 (2010): 425,425-430. Print.

2. Hasegawa, E., et al. "Ecology of Armillaria Species on Conifers in Japan." Forest Pathology 41 (2011): 429,429-437. Print.

3. Ide, J., et al. "Effects of Discharge Level on the Load of Dissolved and Particulate Components of Steam Nitrogen and Phosphorus from a Small Afforested Watershed of Japanese Cypress (Chamaecyparis Obtusa)." The Japanese Forest Society and Springer 12 (2007): 45,45-56. Print.

4. Miyamoto, Asako, and Makoto Sano. "The Influence of Forest Management on Landscape Structure in the Cool-Temperate Forest Region of Central Japan." Landscape and Urban Planning 86 (2008): 248,248-256. Print.

5. Noguchi, Mahoko, and Toshiya Yoshida. "Tree Regeneration in Partially Cut Conifer-Hardwood Mixed Forests in Northern Japan: Roles of Establishment Substrate and Dwarf Bamboo." Forest Ecology and Management 190 (2004): 335,335-344. Print.

6. Resco de Dios, Victor, Toshiya Yoshida, and Yoko Iga. "Effects of Topsoil Removal by Soil-Scarification on Regeneration Dynamics of Mixed Forests in HOkkaido, Northern Japan." Forest Ecology and Management 215 (2005): 138,138-148. Print.

7. Shoyama, Kikuko. "Reforestation of Abandoned Pasture on Hokkaido, Northern Japan: Effect of Plantations on the Recovery of Conifer-Broadleaved Mixed Forest." International Consortium of Landscape Ecology and Ecological Engineering and Springer 4 (2008): 11,11-23. Print.

Chapter 5

Arbor Day

For other uses, see Arbor Day (disambiguation).

Arbor Day (or **Arbour**; from the Latin *arbor*, meaning tree) is a holiday in which individuals and groups are encouraged to plant and care for trees. Today, many countries observe such a holiday. Though usually observed in the spring, the date varies, depending on climate and suitable planting season.

5.1 Origins

The naturalist Miguel Herrero Uceda at the monument to the first Arbor Day in the world, Villanueva de la Sierra (Spain) 1805.

5.1.1 First Arbor Day in the world

The Spanish village of Mondoñedo held the first documented arbor plantation festival in the world organized by its mayor in 1594. The place remains as Alameda de los Remedios and it is still planted with lime and horsenut trees. A humble granite and a bronze plate recalls the event. Additionally the small Spanish village of Villanueva de la Sierra is the town where was held the first modern Arbor Day, an initiative launched in 1805, by the local priest with the enthusiastic support of the entire population.[1]

> While Napoleon was ravaging Europe with his ambition in this village in the Sierra de Gata lived a priest, don Ramón Vacas Roxo, which, according to the chronicles, "convinced of the importance of trees for health, hygiene, decoration, nature, environment and customs, decides to plant trees and give a festive air. The festival began on Carnival Tuesday with the ringing of two bells of the church, and the Middle and the Big. After the Mass, and even coated with church ornaments, don Ramón, accompanied by clergies, teachers and a large number of neighbours, planted the first tree, a poplar, in the place known as Valley of the Ejido. Tree plantations continued by Arroyada and Fuente de la Mora. Afterwards, there was a feast, and did not miss the dance. The party and plantations lasted three days. He drafted a manifesto in defence of the trees that was sent to surrounding towns to spread the love and respect for nature, and also he advised to make tree plantations in their localities.
> —Miguel Herrero Uceda, Arbor Day

5.1.2 First American Arbor Day

The first American Arbor Day was originated in Nebraska City, Nebraska, United States by J. Sterling Morton. On April 10, 1872, an estimated one million trees were planted in Nebraska.*[1]

Birdsey Northrop of Connecticut was responsible for globalizing it when he visited Japan in 1883 and delivered his Arbor Day and Village Improvement message. In that same year, the American Forestry Association made Northrop the Chairman of the committee to campaign for Arbor Day

Birdsey Northrop

nationwide. He also brought his enthusiasm for Arbor Day to Australia, Canada, and Europe.*[2]

5.1.3   McCreight and Roosevelt

Beginning in 1906, Pennsylvania conservationist Major Israel McCreight of DuBois, Pennsylvania, argued that President Theodore Roosevelt's conservation speeches were limited to businessmen in the lumber industry and recommended a campaign of youth education and a national policy on conservation education.*[3] McCreight urged President Roosevelt to make a public statement to school children about trees and the destruction of American forests. Conservationist Gifford Pinchot, Chief of the United States Forest Service, embraced McCreight's recommendations and asked the President to speak to the public school children of the United States about conservation. On April 15, 1907, Roosevelt issued an "Arbor Day Proclamation to the School Children of the United States" *[4] about the importance of trees and that forestry deserves to be taught in U.S. schools. Pinchot wrote McCreight, "we shall all be indebted to you for having made the suggestion." *[5]

Arbor Day in Algeria

5.2   Around the world

5.2.1   Australia

National Schools Tree Day is held on the last Friday of July for schools and National Tree Day the last Sunday in July throughout Australia. Many states have Arbor Day although only Victoria has Arbor Week, which was suggested by Premier Dick Hamer in the 1980s. Arbor Day has been observed in Australia since 20 June 1889.

5.2.2   Belgium

International Day of Treeplanting is celebrated in Flanders on or around 21 March as a theme-day/educational-day/observance, not as public holidays. Tree planting is sometimes combined with awareness campaigns of the fight against cancer: *Kom Op Tegen Kanker*.

5.2.3   Brazil

The Arbor Day (Dia da Árvore) is celebrated on September 21. It's not a national holiday. However, schools nationwide celebrate this day with environment-related activities, namely tree planting.

5.2.4   British Virgin Islands

Arbour Day is celebrated on November 22. It is sponsored by the National Parks Trust of the Virgin Islands. Activities include an annual national Arbour Day Poetry Competition and tree planting ceremonies throughout the territory.

5.2.5 Cambodia

Cambodia celebrates arbor day on 9 July with a tree planting ceremony attended by the king.[6]

5.2.6 Canada

Founded by Don Clark of Schomberg, Ontario for his wife Margret Clark in 1906. In Canada, Maple Leaf Day falls on the last Wednesday in September during National Forest Week. Ontario celebrates Arbor Week from the last Friday in April to the first Sunday in May. Nova Scotia celebrates Arbor Day on the Thursday during National Forest Week, which is the first full week in May. Prince Edward Island celebrates Arbor Day on the 3rd Friday in May during Arbor Week.[7]

5.2.7 Central African Republic

National Tree Planting Day is on July 20.

5.2.8 China

In 1981, the fourth session of the Fifth National People's Congress of the People's Republic of China adopted the Resolution on the Unfolding of a Nationwide Voluntary Tree-planting Campaign. This resolution established the Arbor Day (Chinese: 植树节) and stipulated that every able-bodied citizen between the ages of 11 and 60 should plant three to five trees per year or do the equivalent amount of work in seedling, cultivation, tree tending or other services. Supporting documentation instructs all units to report population statistics to the local afforestation committees as the basis for workload allocation. Moreover, those failing to do their duty are expected to make up planting requirements, provide funds equivalent to the value of labor required or pay heavy fines. Therefore, the tree-planting campaign is actually compulsory, or at least obligatory (that is, an obligation to the community). The "voluntary" in the title referred to the fact that the tree-planters would "volunteer" their labour. The People's Republic of China celebrates Arbor Day on March 12, a day founded by Lin Daoyang, continue to use following the date of Arbor Day of Republic of China.

5.2.9 Costa Rica

"Día del Árbol" is on June 15.

5.2.10 Egypt

Arbor Day is on January 15.

5.2.11 Germany

Arbor Day ("Tag des Baumes") is on April 25. First celebration was in 1952.

5.2.12 India

Van Mahotsav is an annual pan-Indian tree planting festival, occupying a week in the month of July. During this event millions of trees are planted. It was initiated in 1950 by K. M. Munshi, the then Union Minister for Agriculture and Food to create an enthusiasm in the mind of the populace for the conservation of forests and planting of trees.

The name Van Mahotsava (the festival of trees) originated in July 1947 after a successful tree-planting drive was undertaken in Delhi, in which national leaders like Jawaharlal Nehru, Dr Rajendra Prasad and Abul Kalam Azad participated. Paryawaran Sachetak Samiti, a leading environmental organization conducts mass events & concrete activities on this special day celebration each year. The week was simultaneously celebrated in a number of states in the country.

5.2.13 Iran

In Iran it is known as National Tree Planting Day. By Solar Hijri calendar, it is on the 15th day of month Esfand which usually corresponds with March 5.
This day is the first day of the Natural Recyclable Resources week (March 5 to 12).
This is the time in which the saplings of the all kinds in terms of different climates of different parts of Iran would be shared among the people. They also are going to be taught the ways of planting trees.[8]

5.2.14 Israel

The Jewish holiday Tu Bishvat, the new year for trees, is on the 15th day of the month of Shvat, which usually falls in January or February. Originally based on the date used to calculate the age of fruit trees for tithing as mandated in Leviticus 19:23–25, the holiday now is most often observed by planting trees, or raising money to plant trees.[9] Tu Bishvat is a semi official holiday in Israel, schools are open but Hebrew speaking schools will often go on tree planting excursions.

Tu Bishvat, Israel

5.2.15 Japan

Japan celebrates a similarly themed Greenery Day, held on May 4. Although it has a similar theme to Arbor Day, its roots lie in celebration of the birthday of Emperor Hirohito.

5.2.16 Kenya

National Tree Planting Day is on April 21. Often people plant palm trees and coconut trees along the Indian Ocean that borders the East coast of Kenya.

5.2.17 Lesotho

National Tree Planting Day is on March 21.

5.2.18 Luxembourg

National Tree Planting Day is in November since 1991. It is organized by natur&ëmwelt.

5.2.19 Republic of Congo

National Tree Planting Day is on November 6.

5.2.20 Republic of Macedonia

Having in mind the bad condition of the forest fund, and in particular the catastrophic wildfires which occurred in the summer of 2007, a citizen's initiative for afforestation was started in the Republic of Macedonia. The campaign by the name 'Tree Day-Plant Your Future' was first organized on 12 March 2008, when an official non-working day was declared and more than 150,000 Macedonians planted 2 million trees in one day (symbolically, one for each citizen). Six million more were planted in November the same year, and another 12,5 million trees in 2009. This has been established as a tradition and takes place every year.

5.2.21 Malawi

National Tree Planting Day is on the 2nd Monday of December.

5.2.22 Mexico

President Enrique Peña Nieto plants a tree in Balleza, Chihuahua to commemorate the Día del Árbol *2013.*

The *Día del Árbol* was established in Mexico in 1959 with President Adolfo López Mateos issuing a decree that it should be observed on the 2nd Thursday of July.[10]

5.2.23 Mongolia

National Tree Planting Day is on the 2nd Saturday of May and October. It is first National Tree Planting Day was celebrated on 2010-05-08

5.2.24 Namibia

Its first Arbor Day was celebrated on October 8 2004.*[11]

5.2.25 Netherlands

Since conference and of the Food and Agriculture Organization's publication *World Festival of Trees*, and a resolution of the United Nations in 1954: "The Conference, recognising the need of arousing mass consciousness of the aesthetic, physical and economic value of trees, recommends a World Festival of Trees to be celebrated annually in each member country on a date suited to local conditions"; it has been adopted by the Netherlands. In 1957, the National Committee Day of Planting Trees/Foundation of National Festival of Trees (*Nationale Boomplantdag/Nationale Boomfeestdag*) was created.

On the third Wednesday in March each year (near the spring equinox), three quarters of Dutch schoolchildren aged 10/11 and Dutch celebrities plant trees. Stichting Nationale Boomfeestdag organizes all the activities in the Netherlands for this day. Some municipalities however plant the trees around 21 September because of the planting season.*[12]

In 2007, the 50th anniversary was celebrated with special golden jubilee-activities.

5.2.26 New Zealand

New Zealand's first Arbor Day planting was in Greytown in the Wairarapa on 3 July 1890. The first official celebration will take place in Wellington in August 2012, with the planting of pohutukawa and Norfolk pines along Thorndon Esplanade.

Born in 1855, Dr Leonard Cockayne (generally recognised as the greatest botanist who has lived, worked, and died in New Zealand) worked extensively on native plants throughout New Zealand and wrote many notable botanical texts. Even as early as the 1920s he held a vision for school students of New Zealand to be involved in planting native trees and plants in their school grounds. This vision bore fruit and schools in New Zealand have long planted native trees on **Arbor Day**.

Since 1977, New Zealand has celebrated Arbor Day on June 5, which is also World Environment Day, prior to then Arbor Day, in New Zealand, was celebrated on August 4 – which is rather late in the year for tree planting in New Zealand hence the date change.

What the Department of Conservation (DOC) does for Arbor Day: Many of DOC's Arbor Day activities focus on ecological restoration projects using native plants to restore habitats that have been damaged or destroyed by humans or invasive pests and weeds. There are great restoration projects underway around New Zealand and many organisations including community groups, landowners, conservation organisations, iwi, volunteers, schools, local businesses, nurseries and councils are involved in them. These projects are part of a vision to protect and restore the indigenous biodiversity.

5.2.27 Niger

Since 1975, Niger has celebrated Arbor Day as part of its Independence Day: 3 August. On this day, aiding the fight against desertification, each Nigerien plants a tree.

5.2.28 Pakistan

National tree plantation day of Pakistan (قومی شجر کاری دن) is celebrated on 18 August.*[13]

5.2.29 Philippines

Arbor Day in the Philippines has been institutionalized to be observed every June 25 throughout the nation by planting trees and ornamental plants and other forms of relevant activities. The necessity to promote a healthier ecosystem for the people through the rehabilitation and regreening of the environment was stressed in Proclamation No. 643 that amended Proclamation No. 396 of June 2, 2003. Proclamation No. 396 enjoined the "active participation of all government agencies, including government-owned and controlled corporations, private sector, schools, civil society groups and the citizenry in tree planting activity and declaring June 25, 2003 as Philippines Arbor Day."

5.2.30 Poland

In Poland, Arbor Day is celebrated since 2002. Each year, on October 10. lots of Polish people plant trees as well as participate in events organized by ecological foundations. Moreover, Polish Forest Inspectorates and schools give special lectures and lead ecological awareness campaigns.

5.2.31 Portugal

Arbor Day is celebrated on March 21. It's not a national holiday but instead schools nationwide celebrate this day with environment-related activities, namely tree planting.

5.2.32   South Africa

Arbor Day was celebrated from 1945 until 2000 in South Africa, when the national government extended it to National Arbor Week, which lasts from 1–7 September. Two trees, one common and one rare, are highlighted to increase public awareness of indigenous trees, while various "greening" activities are undertaken by schools, businesses and other organizations.

5.2.33   South Korea

Main article: Sikmogil

Arbor Day (Sikmogil, □□□) was a public holiday in South Korea on April 5 until 2005. The day is still celebrated, though. On non-leap years, the day coincides with Hansik.

5.2.34   Spain

Planting holm oaks in Pescueza

In Spain is usually held on the International Forest Day on 21 March, but following the 1915 decree spirit that forced to celebrate the Arbor Day throughout Spain, each municipality or collective decides the date for its Arbor Day, usually between February and May. In Villanueva de la Sierra (Extremadura), where the first Arbor Day in the world was held in 1805, it is celebrated, as on that occasion, on Tuesday Carnaval. It is a great day in the local festive calendar.[14]

As an example of commitment to nature, the small town of Pescueza, with only 180 inhabitants, organizes every spring a large plantation of holm oaks, which is called the "Festivalino" promoted by city council, several foundations and citizen participation where several thousand people together repopulates naked lands and regaining life. All

wrapped up in a fun party atmosphere, joy, music and renew.[15]

5.2.35   Sri Lanka

National Tree Planting Day is on November 15.

5.2.36   Taiwan

Arbor Day (植樹節) has been a traditional holiday in the **Republic of China** since 1927. In 1914, the founder of the agricultural college at Nanking University suggested to the now-defunct Ministry of Agriculture and Forestry that China should imitate the practice in the United States of Arbor Day. The holiday would be held the same day as the Qingming Festival. However, for unknown reasons, the suggestion was not made through the formal process, so nothing came from this original request. After the successful conclusion of the Northern Expedition, the now-defunct Ministry of Agriculture and Minerals formally petitioned the Executive Yuan to establish Arbor Day to commemorate the passing of Dr. Sun Yat-sen, the Father of Modern China. He had been a major advocate of afforestation in his life, because it would increase people's livelihoods. The Executive Yuan approved Arbor Day in the spirit of Dr. Sun that year and has since been celebrated on March 12 for this purpose.

5.2.37   Tanzania

National Tree Planting Day is on April 1

5.2.38   Uganda

National Tree Planting Day is on March 24.

5.2.39   United Kingdom

First mounted in 1975, National Tree Week is a celebration of the start of the winter tree planting season. Around a million trees are planted each year by schools, community organizations and local authorities.

5.2.40   United States

Arbor Day was founded in 1872 by Julius Sterling Morton in Nebraska City, Nebraska. By the 1920s, each state in the United States had passed public laws that stipulated a

Arbor Day community festival in Rochester, Minnesota

certain day to be Arbor Day or Arbor and Bird Day observance.

National Arbor Day is celebrated every year on the last Friday in April; in Nebraska, it is a civic holiday. Each state celebrates its own state holiday. The customary observance is to plant a tree. On the first Arbor Day, April 10, 1872, an estimated one million trees were planted.*[1]

5.2.41 Venezuela

Venezuela recognizes "Día del Arbol" on the last Sunday of May.

5.3 See also

- Arbor Day Foundation (USA)

- Earth Day

- Greenery Day (Japan)

- International Day of Forests

- National Public Lands Day (USA)

- Timeline of environmental events

- Tu Bishvat

- World Water Day

5.4 References

[1] "The History of Arbor Day," Arborday.org. Accessed: April 26, 2013.

[2] *Birdsey Grant Northrop* (PDF), retrieved 2009-04-25

[3] M.I. McCreight, Theodore Roosevelt and Conservation Why: A Thirty-Four Year Moratorium on Unpublished Records (1940), Historical Society of Western Pennsylvania, at p.12, Hereinafter cited "Theodore Roosevelt and Conservation Why".

[4] Arbor Day Proclamation to the School Children of the United States

[5] "Theodore Roosevelt and Conservation Why"

[6] "Cambodian King Attends the Celebration of Annual Arbor Day (July 9)". *Agenche Kampuchea Presse*. Agenche Kampuchea Presse. 9 July 2012. Retrieved 11 October 2015.

[7] "Arbor Day Around The World". The Arbor Day Foundation. 20 July 2010

[8] "Tree Planting Day". *tebyan.net*. Tebyan Cultural and Information Center. Retrieved 3 March 2014.

[9] *Judaism 101: Tu B'Shevat*. Accessed August 20, 2007.

[10] "Hoy, Día del Árbol en México". Azteca Noticias. Retrieved 6 October 2015.

[11] "Arbor Day Around The World". Arbor day foundation. Archived from the original on 18 September 2008. Retrieved 2008-09-06.

[12] Boomfeestdag http://www.boomfeestdag.nl/ the organisations address is Spoorlaan 444 5038 CH TILBURG

[13] 18 August declared as NTPD

[14] The oldest environmentalist festival in the world was celebrated in Villanueva. Sierra de Gata News. February 26, 2014

[15] Herrero Uceda, Miguel y Elisa: *Mi Extremadura*. 2011, pages 147-148

5.5 External links

- International Arbor Days

- History of Arbor Day

- Arbor Day Lesson Plans for the Classroom

- National Arbor Day Foundation

- State Arbor Days and state trees

-

- *Arbor Day Leaves – A Complete Programme For Arbor Day Observance, Including Readings, Recitations, Music, and General Information* at Project Gutenberg

Chapter 6

Armenia Tree Project

Armenia Tree Project (**ATP**) is a non-profit organization based in Watertown, Massachusetts, United States, and Yerevan, Armenia, founded in 1994 by Carolyn Mugar to promote Armenia's socioeconomic development through reforestation. Since its founding, the organization has planted more than 4.5 million trees in communities throughout Armenia.[3][4]

The organization has a full-time staff of 80, of whom 75 are employed in Armenia. Its Yerevan branch manages three state-of-the-art tree nurseries, two environmental education centers, and partners with families to create tree-based small business opportunities.[5] Its major program initiatives include planting trees at urban and rural sites, environmental education and advocacy, community development and poverty reduction.[6]

6.1 Environmental challenge

When Carolyn Mugar, from Boston, visited Armenia in 1992, the country had been impoverished by an energy embargo imposed during the Nagorno-Karabakh War. Armenians had previously depended upon natural gas for 90 percent of their energy needs, but their supply had been cut off by the embargo. Deforestation was particularly severe during the early 1990s, because many Armenians had only their trees as a fuel source during the winter. This condition raised a concern about whether land formerly protected by forest cover would become desert. A study in 2005 estimated Armenia's forest cover at 11.2 percent of its total land area, dropping to 8.2 percent by 2000.[7][3] In 2012, the ATP reported the country's forest cover down to only 7 percent. [8]

In 1994, Carolyn Mugar established the Armenia Tree Project to address the environmental and economic disaster of Armenia's dwindling forests. The ATP was organized as a subsidiary of the Armenian Assembly of America, which continues to provide administrative assistance. Since its founding, the ATP has planted over 4.5 mil-

lion trees throughout Armenia. As of 2014, the organization was operating three tree nurseries, providing full-time employment for 45 people, and fruit trees planted by its projects were producing an estimated harvest of over 300,000 pounds annually.[4]

6.2 ATP programs

The organization's mission emphasizes the use of trees to promote economic self-sufficiency, improving the Armenian standard of living while protecting the environment.[9] Its urban and community tree planting programs work with cities and local neighborhoods to replant in public spaces such as in parks, school grounds and other public properties. In rural areas, farmers grow seedlings in their backyards for tree planting projects in northern Armenia.[6]

In its environmental education and advocacy programs, ATP teaches the value of living in a healthy environment. The organization is seeking approval by the Ministry of Education to present an environmental studies curriculum for schools. Its poverty reduction and community development efforts direct funding to employ community residents in tree planting, and teach families to grow and tend seedlings in backyard nursery pots.[6]

Building Bridges is an online program created by ATP for the children of Armenia. It allows them to explore their environmental heritage and play games where they plant their own virtual trees.[10]

6.3 Engergy Globe Award

In 2008, the ATP's tree nursery micro-enterprise program, "Plant an Idea, Plant a Tree", was recognized as the national winner of the Energy Globe Award for Sustainability.[11] Its nursery program was selected for the award from 853 environmental projects in 109 countries. Initiated as a pilot project in 2004, the program was designed to mitigate poverty-driven deforestation with support for tree nurseries owned by impoverished families in the Getik River Valley of northern Armenia. It began with 17 families operating tree nurseries in 2004, growing to 400 families by 2008.*[12]

6.4 Volunteer participation

ATP recruits volunteers with assistance from Birthright Armenia / Depi Hayk Foundation. A limited number of volunteer summer positions are available in public relations and outreach,*[13] environmental education,*[14] and the SEEDS program.*[15]*[16] The organization also hires 75 to 100 seasonal workers each year for large-scale reforestation projects.*[4]

6.5 References

[1] "Message From Our Founder". Armenia Tree Project. Retrieved September 26, 2014.

[2] "Armenia Tree Project Staff". Armenia Tree Project. Retrieved 26 September 2014.

[3] Teodora Gaydarova (April 28, 2014). "Green Gold: Rebuilding Armenia's Forests". *The Armenite.*

[4] Aram Arkun (October 3, 2014). "Armenian Tree Project Celebrates 20th Anniversary". *The Armenian Mirror-Spectator.*

[5] "Mission". Armenia Tree Project. Retrieved September 24, 2014.

[6] "What We Do". Armenia Tree Project. Retrieved 10 August 2014.

[7] Dr. Svetlana Aslanyan (2012). "Armenia: Undermining the environment" (PDF). *Social Watch* (Center for the Development of Civil Society). p. 65.

[8] "The Threat". Armenia Tree Project. Retrieved 10 August 2014.

[9] "Home Page". Armenia Tree Project. Retrieved 10 August 2014.

[10] "Building Bridges". Armenia Tree Project. Retrieved 10 August 2014.

[11] "National ENERGY GLOBE Award Armenia". *Energy-Globe Award Portal.* Retrieved September 26, 2014.

[12] ATP news release (June 11, 2008). "Armenia tree project micro-enterprise program recognized as national winner of Energy Globe Award for Sustainability". *World Wire* (Environmental News Services).

[13] "Public Relations and Outreach" (PDF). Armenia Tree Project.

[14] "Environmental Education" (PDF). Armenida Tree Project.

[15] "SEEDS Program" (PDF). Armenia Tree Project.

[16] "Volunteer". Retrieved 10 August 2014.

6.6 External links

- Official site

- Recent achievements and awards

- *Trees for Life*, a documentary

- ATP's Building Bridges Program: Connecting Diaspora Armenian Students with Their Environmental Heritage

- Armenia Tree Project's Facebook page

Chapter 7

The Big Tree Plant

Launched in December 2010, **The Big Tree Plant** is a Government-sponsored campaign in England to promote the planting of one million trees in neighbourhoods where people live and work.[*][1][*][2] The campaign will run over four years from 2011 to 2015, and is the first such initiative since *Plant A Tree In '73*.[*][1][*][2]

7.1 Background

The campaign aims to halt the ongoing decline in urban and semi-urban tree planting in England.[*][1] The decline was highlighted by a survey of urban trees in England carried out in 2005 (published as the report *Trees in Towns II* in 2008), which found that there had been a 'big reduction' in urban tree planting (compared to a similar 1992 survey) leading to an 'unsatisfactory age structure' with too few young trees, and which concluded that the issue should be 'urgently addressed'.[*][3] In London a separate 2007 report, *Chainsaw Massacre*, found that there were concerns about planting rates in some boroughs, and that mature broadleaf street trees throughout London were under 'severe threat' due to a mixture of development pressures, reduced expenditure, public apathy and antipathy, and (often unsubstantiated) concerns by insurance companies, solicitors and home-owners over subsidence.[*][4] Both reports also express concern over the practice of planting smaller ornamental species rather than native broadleaf trees such as plane, lime and oak.

7.2 Funding

Funding of £4.2m will be made available for community, civic and other non-profit groups from April 2011.[*][5] In addition to covering planting costs, grants can be used for related purposes such as community involvement, site surveys and the provision of expert advice.[*][2] £4m of the funding is being provided by the Forestry Commission[*][2] through 'efficiency savings and reprioritisation',[*][6] while the remaining £200,000 will come from the existing London Tree and Woodland Community Grant.[*][7] The independently chaired Big Tree Plant Grants Panel will include representatives from civil society organisations, DEFRA, and the Forestry Commission, and will meet each spring and summer to award funds.[*][8]

In advance of the main funding, Keep Britain Tidy - one of the partners supporting the initiative - has already invited applications for planting kits from schools in the Government's Eco-School programme.[*][9][*][10]

7.2.1 Criticism

The funding arrangements, specifically the fact that the grants will normally only cover up to 75% of the cost of each scheme (although free labour can be offset against this),[*][11] have been criticised as favouring better-off over deprived communities.[*][12]

7.3 Partners

The DEFRA-led The Big Tree Plant campaign is backed by a number of partners including The Tree Council, Woodland Trust, Trees for Cities, England's 12 Community Forests, the National Forest, BTCV, Civic Voice, Groundwork UK, Keep Britain Tidy, the Local Government Association, the Department for Communities and Local Government and the Forestry Commission.[*][1]

7.4 See also

- Plant A Tree In '73 - a similar campaign in 1973.

- Treeplanting

- Urban forest

7.5   References

[1] The Big Tree Plant: new partnership to plant one million trees, DEFRA, published 2010-12-02, accessed 2010-12-09

[2] Aims of The Big Tree Plant, Forestry Commission, accessed 2010-12-09

[3] Trees in Towns II, Department for Communities and Local Government, published 2008-02-18, accessed 2010-12-09

[4] Chainsaw massacre: A review of London's street trees Greater London Authority, published 2007-05-01, accessed 2010-12-09

[5] Funding and grants The Big Tree Plant, accessed 2010-12-09

[6] The Big Tree Plant launched, Horticulture Week, published 2010-12-02, accessed 2010-12-09

[7] Government launches £4.2m urban tree-planting plan, Manchester Wired, published 2010-12-02, accessed 2010-12-09

[8] Grants Panel, Forestry Commission, accessed 2010-12-09

[9] The Big Tree Plant - order your free tree planting kit, Keep Britain Tidy, accessed 2010-12-09

[10] The Big Tree Plant, Keep Britain Tidy, published 2010-12-06, accessed 2010-12-09

[11] The Big Tree Plant funding scheme, The Big Tree Plant, accessed 2010-12-09

[12] A million trees for England: but who gets them?, The Guardian, published 2010-12-02, accessed 2010-12-09

7.6   External links

- The Big Tree Plant - official site

- The benefits of urban trees - Trees for Cities

Chapter 8

Billion Tree Campaign

The **Billion Tree Campaign** was launched in 2006 by the United Nations Environment Programme (UNEP) as a response to the challenges of global warming, as well as to a wider array of sustainability challenges, from water supply to biodiversity loss.*[1] Its initial target was the planting of one billion trees in 2007.*[2] One year later, in 2008, the campaign's objective was raised to 7 billion trees – a target to be met by the climate change conference that was held in Copenhagen, Denmark in December 2009. Three months before the conference, the 7 billion planted trees mark had been surpassed. In December 2011, after more than 12 billion trees had been planted, UNEP formally handed management of the program over to the not-for-profit Plant-for-the-Planet Foundation, based in Munich, Germany.*[3]

8.1 Inspiration

The Billion Tree Campaign was inspired by Nobel Peace Prize Laureate Wangari Maathai, founder of the Green Belt Movement. When an executive in the United States told Professor Maathai their corporation was planning to plant a million trees, her response was: "That's great, but what we really need is to plant a billion trees." *[4] The campaign was carried out under the patronage of Prince Albert II of Monaco.*[5]

8.2 Response

Under UNEP's leadership and through proactive advocacy by the patrons and the partners, the Billion Tree Campaign catalysed tree planting action on all continents. The billionth tree, an African Olive, was planted in Ethiopia in November 2007.*[6] The two billionth tree took root as part of the United Nation's World Food Programme agroforestry initiative. The campaign's target was then raised to seven billion trees.*[7] In 2009, UNEP mobilized action across the globe through the Twitter for Trees campaign on www.twitter.com/UNEPandYou. The initiative was driven by a simple yet powerful idea: UNEP pledged to plant one tree to feed into the Billion Tree Campaign, for every follower who joined from 5 May 2009 to World Environment Day on 5 June 2009. The campaign was a success, with 10,300 people following the UNEPandYou page by World Environment Day weekend.*[8]

The World Organization of the Scouts Movement also planted trees under the campaign in line with its mandate to study and protect Nature across several countries.*[9]

The United Nations Peacekeeping missions also joined the campaign and planted trees with its field missions in East-Timor, Côte d'Ivoire, Darfur, Lebanon, Haiti, Congo, and Liberia amongst others missions.*[10]

8.3 Roll of Honour top 10 countries

- China 2.8 billion
- India 2.1 billion
- Ethiopia 1.6 billion
- Mexico 785 million
- Turkey 716 million
- Nigeria 612 million
- Kenya 455 million
- Peru 246 million
- Myanmar 191 million
- Cuba 137 million

"Around a flowering tree, one finds many insects." African saying

December 2011, five years since the campaign's launch, the campaign's website www.unep.org/billiontreecampaign proudly registered over 12.5 billion planted trees across 193 countries.

8.4   Handover to the young generation

The Billion Tree Campaign was handed over to the Plant-for-the-Planet Foundation in December 2011, an organisation that has been participating in the Billion Tree Campaign since 2007. The foundation will keep the momentum of the campaign.

8.5   Quote from UNEP

"Looking back over the Billion Tree Campaign's greatest successes, what is most remarkable is not its scale, but its spread. People from all around the world have enthusiastically joined the campaign and planted trees in their own communities." – Achim Steiner, UNEP Executive Director.

8.6   References

[1]   "Commit to Action – Join the Billion Tree Campaign!".

[2]   "Plant for the Planet".

[3]   UNEP. (7 December 2011). "UNEP Billion Tree Campaign Hands Over to the Young People of the Plant-for-the-Planet Foundation." Accessed: 26 September 2013.

[4]   "THE GREEN BELT MOVEMENT".

[5]   "United Nations Billion Tree Campaign Tree Planting Partner".

[6]   "UN Billion Tree Campaign".

[7]   "UNEP Billion Tree Campaign Hands Over to the Young People of the Plant-for-the-Planet Foundation".

[8]   "PLANET EARTH NEEDS ATTENTION".

[9]   "Your Planet Needs You! Unite to Combat Climate Change on World Environment Day".

[10]   "Tree Planting Campaign Hits Four Billion Mark".

8.7   External links

- Official website - Plant-for-the-Planet Foundation, Munich, Germany

- Plant-for-the-Planet Foundation. (7 December 2011). "UNEP Hands Over Billion Tree Campaign"

Chapter 9

Biosequestration

Flowering Corymbia ficifolia, Austins Ferry, Tasmania, Australia.

Biosequestration is the capture and storage of the atmospheric greenhouse gas carbon dioxide by biological processes.

This may be by increased photosynthesis (through practices such as reforestation / preventing deforestation and genetic engineering); by enhanced soil carbon trapping in agriculture; or by the use of algal bio sequestration (see algae bioreactor) to absorb the carbon dioxide emissions from coal, petroleum (oil) or natural gas-fired electricity generation.

Biosequestration as a natural process has occurred in the past, and was responsible for the formation of the extensive coal and oil deposits which are now being burned. It is a key policy concept in the climate change mitigation debate.[*][1]

It does not generally refer to the sequestering of carbon dioxide in oceans (see carbon sequestration and ocean acidification) or rock formations, depleted oil or gas reservoirs (see oil depletion and peak oil), deep saline aquifers, or deep coal seams (see coal mining) (for all see geosequestration) or through the use of industrial chemical carbon dioxide scrubbing.

9.1 The importance of plants in storing atmospheric carbon dioxide

Kew Gardens Waterlily House. David Iliff, 2008

After water vapour (concentrations of which humans have limited capacity to influence) carbon dioxide is the most abundant and stable greenhouse gas in the atmosphere (methane rapidly reacts to form water vapour and carbon dioxide). Atmospheric carbon dioxide has increased from about 280 ppm in 1750 to 383 ppm in 2007 and is increasing at an average rate of 2 ppm pr year.[*][2] The world's oceans have previously played an important role in sequestering atmospheric carbon dioxide through solubility and the action of phytoplankton.[*][3] This, and the likely adverse consequences for humans and the biosphere of associated global warming, increases the significance of investigating policy mechanisms for encouraging biosequestration.

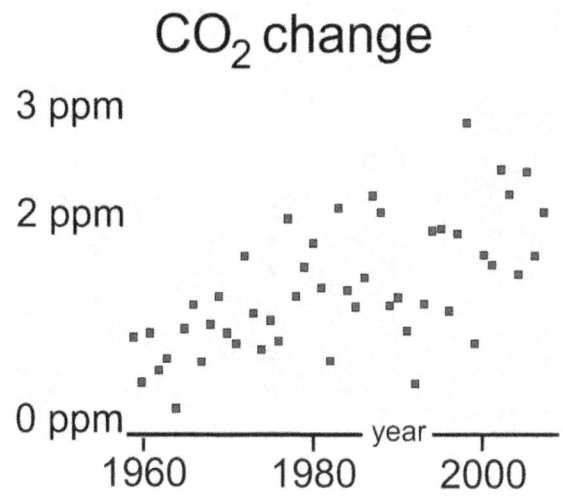

Recent year-to-year increase of atmospheric CO$_2$

9.2 Reforestation, avoided deforestation and LULUCF

Reforestation and reducing deforestation can increase biosequestration in four ways. Pandani (Richea pandanifolia) near Lake Dobson, Mount Field National Park, Tasmania, Australia

The Intergovernmental Panel on Climate Change (IPCC) estimates that the cutting down of forests is now contributing close to 20 per cent of the overall greenhouse gases entering the atmosphere.*[4] Candell and Raupach argue that there are four primary ways in which reforestation and reducing deforestation can increase biosequestration. First, by increasing the volume of existing forest. Second, by increasing the carbon density of existing forests at a stand and landscape scale. Third, by expanding the use of forest products that will sustainably replace fossil-fuel emissions. Fourth, by reducing carbon emissions that are caused from deforestation and degradation.*[5] Land clearing reductions, the majority of the time, create biodiversity benefits in a vast expanse of land regions. Concerns, however,

arise when the density and area of vegetation increases the grazing pressure could also increase in other areas, causing land degradation.*[6]

A recent report by the Australian CSIRO found that forestry and forest-related options are the most significant and most easily achieved carbon sink making up 105 Mt per year CO$_2$-e or about 75 per cent of the total figure attainable for the Australian state of Queensland from 2010-2050. Among the forestry options, the CSIRO report announced, forestry with the primary aim of carbon storage (called carbon forestry) clearly has the highest attainable carbon storage capacity (77 Mt CO$_2$-e/yr) and is one of the easiest options to implement compared with biodiversity plantings, pre-1990 eucalypts, post 1990 plantations and managed regrowth.*[7] Legal strategies to encourage this form of biosequestration include permanent protection of forests in National Parks or on the World Heritage List, properly funded management and bans on use of rainforest timbers and inefficient uses such as woodchipping old growth forest.*[8]

As a result of lobbying by the developing country caucus (or Group of 77) in the United Nations (associated with the United Nations Conference on Environment and Development (UNCED) in Rio de Janeiro, the non-legally binding Forest Principles were established in 1992. These linked the problem of deforestation to third world debt and inadequate technology transfer and stated that the "agreed full incremental cost of achieving benefits associated with forest conservation...should be equitably shared by the international community" (para1(b)).*[9] Subsequently the Group of 77 argued in the 1995 *Intergovernmental Panel on Forests* (IPF) and then the 2001 *Intergovernmental Forum on Forests* (IFF), for affordable access to environmentally sound technologies without the stringency of intellectual property rights; while developed states there rejected demands for a forests fund.*[10] The expert group created under the United Nations Forum on Forests (UNFF) reported in 2004, but in 2007 developed nations again vetoed language in the principles of the final text which might confirm their legal responsibility under international law to supply finance and environmentally sound technologies to the developing world.*[11]

In December 2007, after a two year debate on a proposal from Papua New Guinea and Costa Rica, state parties to the United Nations Framework Convention on Climate Change (FCCC) agreed to explore ways of reducing emissions from deforestation and to enhance forest carbon stocks in developing nations.*[12] The underlying idea is that developing nations should be financially compensated if they succeed in reducing their levels of deforestation (through valuing the carbon that is stored in forests); a concept termed 'avoided deforestation (AD) or, REDD if broadened to include reducing forest degradation (see Reducing emissions

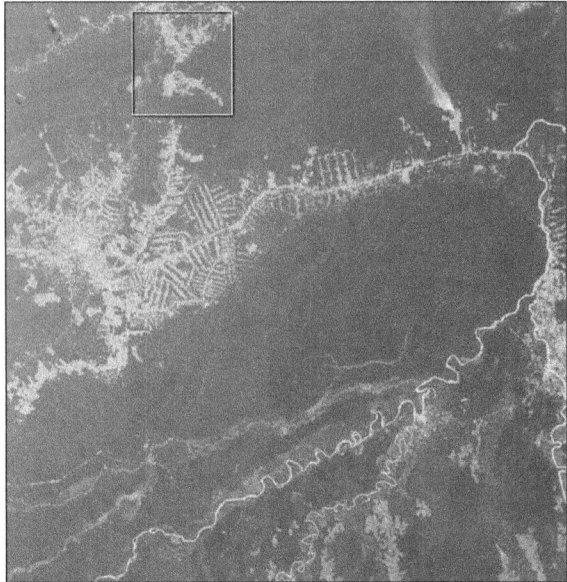

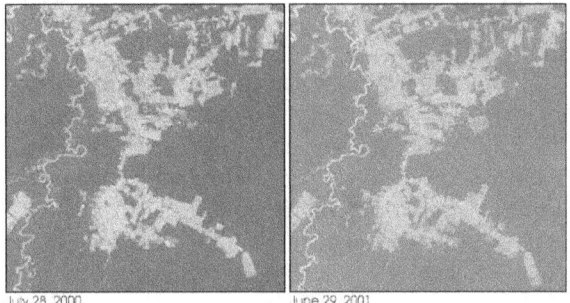

Settlement and deforestation surrounding the Brazilian town of Rio Branco are seen here in the striking "herring bone" deforestation patterns that cut through the rainforest. NASA, 2008.

NASA Earth Observatory, 2009. Deforestation in Malaysian Borneo.

from deforestation and forest degradation). Under the free market model advocated by the countries who have formed the *Coalition of Rainforest Nations*, developing nations with rainforests would sell carbon sink credits under a free market system to Kyoto Protocol Annex I states who have exceeded their emissions allowance.[13] Brazil (the state with the largest area of tropical rainforest) however, opposes including avoided deforestation in a carbon trading mechanism and instead favors creation of a multilateral development assistance fund created from donations by developed states.[13] For REDD to be successful science and regulatory infrastructure related to forests will need to increase so nations may inventory all their forest carbon, show that they can control land use at the local level and prove that their emissions are declining.[14]

Subsequent to the initial donor nation response, the UN established REDD Plus, or REDD+, expanding the original program's scope to include increasing forest cover through both reforestation and the planting of new forest cover, as well as promoting sustainable forest resource manage-

ment.[15]

The United Nations Framework Convention on Climate Change (UNFCCC) Article 4(1)(a) requires all Parties to "develop, periodically update, publish and make available to the Conference of the Parties" as well as "national inventories of anthropogenic emissions by sources" "removals by sinks of all greenhouse gases not controlled by the Montreal Protocol." Under the UNFCCC reporting guidelines, human-induced greenhouse emissions must be reported in six sectors: energy (including stationary energy and transport); industrial processes; solvent and other product use; agriculture; waste; and *land use, land use change and forestry* (LULUCF).[16] The rules governing accounting and reporting of greenhouse gas emissions from LULUCF under the Kyoto Protocol are contained in several decisions of the Conference of Parties under the UNFCCC and LULUCF has been the subject of two major reports by the Intergovernmental Panel on Climate Change (IPCC).[17] The Kyoto Protocol article 3.3 thus requires mandatory LULUCF accounting for afforestation (no forest for last 50 years), reforestation (no forest on 31 December 1989) and deforestation, as well as (in the first commitment period) under article 3.4 voluntary accounting for cropland management, grazing land management, revegetation and forest management (if not already accounted under article 3.3).[18]

As an example, the *Australian National Greenhouse Gas Inventory* (NGGI) prepared in compliance with these requirements indicates that the energy sector accounts for 69 per cent of Australia's emissions, agriculture 16 per cent and LULUCF six per cent. Since 1990, however, emissions from the energy sector have increased 35 per cent (stationary energy up 43% and transport up 23%). By comparison, emissions from LULUCF have fallen by 73%.[19] However, questions have been raised by Andrew Macintosh

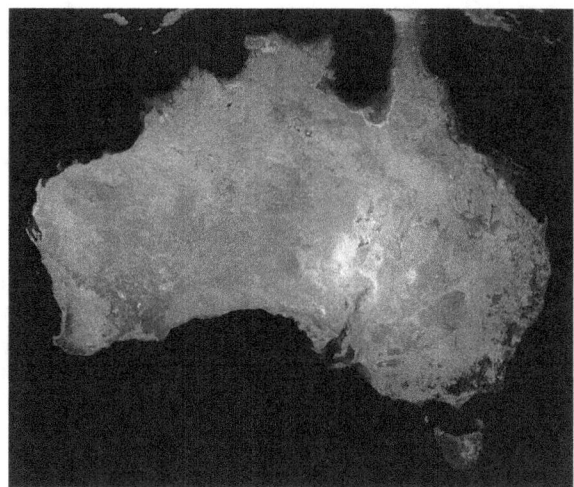

Continent of Australia from space. Australia is a major producer of fossil fuels and has significant problems with deforestation.

Deforestation in Haiti. NASA, 2008.

about the veracity of the estimates of emissions from the LULUCF sector because of discrepancies between the Australian Federal and Queensland Governments' land clearing data. Data published by the *Statewide Landcover and Trees Study* (SLATS) in Queensland, for example, show that the total amount of land clearing in Queensland identified under SLATS between 1989/90 and 2000/01 is approximately 50 per cent higher than the amount estimated by the Australian Federal Government's *National Carbon Accounting System* (NCAS) between 1990 and 2001.[20]

Satellite imaging has become crucial in obtaining data on levels of deforestation and reforestation. Landsat satellite data, for example, has been used to map tropical deforestation as part of NASA's Landsat *Pathfinder Humid Tropical Deforestation Project*, a collaborative effort among scientists from the University of Maryland, the University of New Hampshire, and NASA's Goddard Space Flight Cen-

ter. The project yielded deforestation maps for the Amazon Basin, Central Africa, and Southeast Asia for three periods in the 1970s, 1980s, and 1990s.[21]

9.3   Enhanced photosynthesis

Sprekelia formosissima *in Tasmania, Australia.*

Biosequestration may be enhanced by improving photosynthetic efficiency by modifying RuBisCO genes in plants to increase the catalytic and/or oxygenation activity of that enzyme.[22]

One such research area involves increasing the Earth's proportion of C4 carbon fixation photosynthetic plants. C4 plants represent about 5% of Earth's plant biomass and 1% of its known plant species,[23] but account for around 30% of terrestrial carbon fixation.[24] In leaves of C3 plants, captured photons of solar energy undergo photosynthesis which assimilates carbon into carbohydrates (triosephosphates) in the chloroplasts of the mesophyll cells. The primary CO_2 fixation step is catalysed by ribulose-1,5-bisphosphate carboxylase/oxygenase (Rubisco) which reacts with O2 leading to photorespiration that protects photosynthesis from photoinhibition but wastes 50% of potentially fixed carbon.[25] The C4 photosynthetic pathway, however, concentrates CO_2 at the site of the reaction of Rubisco, thereby reducing the biosequestration-inhibiting photorespiration.[26] A new frontier in crop science consists of attempts to genetically engineer C3 staple food crops (such as wheat, barley, soybeans, potatoes and rice) with the "turbo-charged" photosynthetic apparatus of C4 plants.[27]

carbon from biomass as a cheaper mitigation option than geosequestration by CO_2 capture and storage.[31]

9.5 Improved agricultural and farming practices

Zero-till farming practices occur where there is much mulching but ploughing is not used, so that the carbon-rich organic matter in soil is not exposed to atmospheric oxygen, or to the leaching and erosion effects of rainfall. Ceasing ploughing has been alleged to encourage more ants to become predators of wood-eating (and CO_2 generating) termites, allows weeds to regenerate soils and helps slow water flows over the land.[32]

Shepherds with their sheep.

Soil holds more carbon than vegetation and atmosphere combined, and most soil lies under grazing land.[33][34] Holistic Planned Grazing holds tremendous potential in mitigating global warming, while building soil, increasing biodiversity, and reversing desertification.[35][36] Developed by Allan Savory,[37] it uses fencing and/or herders, to restore grasslands[38][39][40] by carefully planning movements of large herds of livestock to mimic the vast herds found in nature where grazing animals are kept concentrated by pack predators and forced to move on after eating, trampling, and manuring an area, returning only after it has fully recovered. This method of grazing seeks to emulate what occurred during the past 40 million years as the expansion of grass-grazer ecosystems built deep, rich grassland soils, sequestering carbon and cooling the planet.[41]

Dedicated biofuel and biosequestration crops, such as switchgrass (panicum virgatum), are also being developed.[42] It requires from 0.97 to 1.34 GJ fossil energy to produce 1 tonne of switchgrass, compared with 1.99 to 2.66 GJ to produce 1 tonne of corn.[43] Given that switchgrass contains approximately 18.8 GJ/ODT of biomass, the energy output-to-input ratio for the crop can be up to 20:1.[44]

Hakea epiglottis, *Cape Raoul, Tasman Peninsula, Tasmania, Australia.*

9.4 Biochar

Biochar (charcoal created by pyrolysis of biomass) is a potent form of longterm (thousands of years) biosequestration of atmosphereic CO_2 derived from investigation of the extremely fertile Terra preta soils of the Amazon Basin.[28] Placing biochar in soils also improves water quality, increases soil fertility, raises agricultural productivity and reduce pressure on old growth forests.[29] As a method of generating bio-energy with carbon storage Rob Flanagan and the EPRIDA biochar company have developed low-tech cooking stoves for developing nations that can burn agricultural wastes such as rice husks and produce 15% by weight of biochar; while BEST Energies in NSW Australia have spent a decade developing an Agrichar technology that can combust 96 tonnes of dry biomass each day, generating 30-40 tonnes of biochar.[30] A parametric study of biosequestration by Malcolm Fowles at the Open University, indicated that to mitigate global warming, policies should encourage displacement of coal with biomass as a power source for baseload electricity generation if the latter's conversion efficiency rose over 30%, otherwise *biosequestering*

Panicum virgatum *switchgrass, valuable in biofuel production, soil conservation and biosequestration*

Biosequestration can also be enhanced by farmers choosing crops species that produce large numbers of phytoliths. Phytoliths are microscopic spherical shells of silicon that can store carbon for thousands of years.*[45]

9.6   Biosequestration and climate change policy

Biosequestration could be critical to climate change mitigation till cleaner forms of power generation are established. The Nesjavellir Geothermal Power Plant in Þingvellir, Iceland

Industries with large amounts of CO_2 emissions (such as the coal industry) are interested in biosequestration as a means of offsetting their greenhouse gas production.*[46] In Australia, university researchers are engineering algae to produce biofuels (hydrogen and biodiesel oils) and investigating whether this process can be used to *biosequester* carbon. Algae naturally capture sunlight and use its energy to split water into hydrogen, oxygen and oil which can be extracted. Such clean energy production also can be coupled

Windturbines D4 (nearest) to D1 on the Thornton Bank

with desalination using salt-tolerant marine algae to generate fresh water and electricity.*[47]

Many new bioenergy (biofuel) technologies, including cellulosic ethanol biorefineries (using stems and branches of most plants including crop residues such as corn stalks, wheat straw and rice straw) are being promoted because they have the added advantage of biosequestration of CO_2.*[48] The Garnaut Climate Change Review recommends that a carbon price in a carbon emission trading scheme could include a financial incentive for biosequestration processes.*[49] Garnaut recommends the use of algal biosequestration (see algae bioreactor) to absorb the constant stream of carbon dioxide emissions from coal-fired electricity generation and metal smelting until renewable forms of energy, such as solar and wind power, become more established contributors to the grid.*[50] Garnaut, for example, states: "Some algal biosequestration processes could absorb emissions from coal-fired electricity generation and metals smelting." *[51] The United Nations Collaborative Programme on Reducing Emissions from Deforestation and Forest Degradation in Developing Countries (UN-REDD Programme) is a collaboration between FAO,

UNDP and UNEP under which a trust fund established in July 2008 allows donors to pool resources to generate the requisite transfer flow of resources to significantly reduce global emissions from deforestation and forest degradation.*[52] The UK government's Stern Review on the economics of climate change argued that curbing deforestation was a "highly cost-effective way of reducing greenhouse gas emissions".*[53]

James E. Hansen argues that, "An effective way to achieve drawdown [of carbon dioxide] would be to burn biofuels in power plants and capture the CO_2, with the biofuels derived from agricultural or urban wastes or grown on degraded lands using little or no fossil fuel inputs." *[54] Such CO_2 drawdown systems are referred to as Bio-energy with carbon capture and storage, or BECCS. According to a study by Biorecro and the Global CCS Institute, there is currently (as of January 2012) 550 000 tonnes CO_2/year in total BECCS capacity operating, divided between three different facilities.*[55]

Under a 2009 agreement, Loy Yang Power and MBD Energy Ltd will build a pilot Fossil fuel power plant at the Latrobe Valley power station in Australia using biosequestration technology in the form of an algal synthesiser system. Captured CO_2 from the waste exhaust flue gases will be injected into circulating waste water to grow oil-rich algae where sunlight and nutrients will produce heavy oil-laden slurry that can make high grade oil for energy, or stock feed.*[56] Other commercial demonstration projects involving biosequestration of CO_2 at point of emission have begun in Australia.*[57]

9.7 Philosophical basis of biosequestration

The arguments for biosequestration are often shaped in terms of economic theory, yet there is a well-recognised quality of life dimension to this debate.*[58] Biosequestration assists human beings to increase their collective and individual contributions to the essential resources of the biosphere.*[59] The policy case for biosequestration overlaps with principles of ecology, sustainability and sustainable development, as well as biosphere, biodiversity and ecosystem protection, environmental ethics, climate ethics and natural conservation.

Lassen National Park, Kings Creek, USA.

9.8 Barriers to increased global biosequestration

The Garnaut Climate Change Review notes many barriers to increased global biosequestration. "There must be changes in the accounting regimes for greenhouse gases. Investments are required in research, development and commercialisation of superior approaches to biosequestration. Adjustments are required in the regulation of land use. New institutions will need to be developed to coordinate the interests in utilisation of biosequestration opportunities across small business in rural communities. Special efforts will be required to unlock potential in rural communities in developing countries." *[60] Saddler and King have argued that biosequestration and agricultural greenhouse gas emissions should not be handled within a global emissions trading scheme because of difficulties with measuring such emissions, problems in controlling them and the burden that would be placed on numerous small-scale farming operations.*[61] Collett likewise maintains that REDD credits (post-facto payments to developing countries for reducing their deforestation rates below an historical or projected reference rate), simply create a complex market approach to this global public health problem that reduces transparency and accountability when targets are not met and will not be as effective as developed nations voluntarily funding countries to keep their rainforests.*[62]

The World Rainforest Movement has argued that poor developing countries could be pressured to accept reforestation projects under the Kyoto Protocol's Clean Development Mechanism in order to earn foreign exchange simply to pay off the interest on debt to the World Bank.*[63] Tensions also exist over forest management between the sovereignty claims of nations states, arguments about common heritage of mankind and the rights of indigenous peoples and local communities; the Forest Peoples Programme (FPP) arguing the anti-deforestation programs could merely allow financial benefits to flow to national treasuries, privilege would-be corporate forest degraders who manipulate the system by periodically threatening forests, rather than local communities who conserve them.*[64] The success of such projects will also depend on the accuracy of the baseline data and the number of countries involved. Further, it has been argued that if biosequestration is to play a significant role in mitigating anthropogenic climate change then coordinated policies should set a goal of achieving global forest cover to its extent prior to the industrial revolution in the 1800s.*[65]

It has also been argued that the United Nations mechanism for Reducing Emissions from Deforestation and Forest Degradation (REDD) may increase pressure to convert or modify other ecosystems, especially savannahs and wetlands, for food or biofuel, even though those ecosystems also have high carbon sequestration potential. Globally, for example, peatlands cover only 3% of the land surface but store twice the amount of carbon as all the world's forests, whilst mangrove forests and saltmarshes are examples of relatively low-biomass ecosystems with high levels of productivity and carbon sequestration.*[66] Other researchers have argued that REDD is a critical component of an effective global biosequestration strategy that could provide significant benefits, such as the conservation of biodiversity, particularly if it moves away from focusing on protecting forests that are most cost-effective for reducing carbon emissions (such as those in Brazil where agricultural opportunity costs are relatively low, unlike Asia, which has sizeable revenues from oil palm, rubber, rice, and maize). They argue REDD could be varied to allow funding of programs to slow peat degradation in Indonesia and target protection of biodiversity in "hot spot"—areas with high species richness and relatively little remaining forest. Some purchasers, they maintain, of REDD carbon credits, such as multinational corporations or nations, might pay a premium to save imperiled eco-systems or areas with high-profile species.*[67]

9.9   See also

- Bio-energy with carbon capture and storage
- Carbon dioxide removal
- Carbon negative
- Fossil-fuel power station
- Greenhouse gas remediation
- Negative emissions

9.10   References

[1] Garnaut 2008, p. 558 p. 609 defines biosequestration as involving greenhouse gases in general.

[2] Garnaut 2008, p. 33

[3] Raven JA, Falkowski PG (1999). "Oceanic sinks for atmospheric CO_2". *Plant Cell & Environment* **22**: 741–55. doi:10.1046/j.1365-3040.1999.00419.x.

[4] Intergovernmental Panel on Climate Change * The IPCC web site

[5] Canadell JG, Raupach MR (2008). "Managing Forests for Climate Change". *Science* **320** (5882): 1456–7. Bibcode:2008Sci...320.1456C. doi:10.1126/science.1155458. PMID 18556550.

[6] "An analysis of greenhouse gas mitigation and carbon biosequestration opportunities from rural land use" CSIROAugust 2009Website12/4/2013http://www.fcrn.org.uk/sites/default/files/prdz.pdf

[7] CSIRO An Analysis of Greenhouse Gas Mitigation and Carbon Biosequestration Opportunities from Rural Land Use. Canberra. 2009. http://www.csiro.au/resources/carbon-and-rural-land-use-report.html, last accessed 8 October 2009

[8] Diesendorf, Mark (2009). *Climate action: a campaign manual for greenhouse solutions.* Sydney: University of New South Wales Press. p. 116. ISBN 978-1-74223-018-4.

[9] United Nations. Non-Legally Binding Authoritative Statement of Principles for a Global Consensus on the Management, Conservation and Sustainable Development of all Types of Forests. A/CONF.151/6/Rev1. United Nations, Rio de Janeiro. 1992.

[10] Humphreys, David (2006). *Logjam: Deforestation and the Crisis of Global Governance.* London: Earthscan. p. 280. ISBN 1-84407-301-7.

[11] United Nations. Non-Legally Binding Instrument on All Types of Forests. United Nations 22 Oct. 2007. A/C.2/62/L.5.

[12] United Nations. 2007. Reducing emissions from deforestation in developing countries: approaches to stimulate action. http://unfccc.int/files/meetings/cop_13/application/pdf/cp_redd.pdf accessed 10 November 2009.

[13] Humphreys 2008, p. 434

[14] "On the road to REDD". *Nature* **462** (7269): 11. November 2009. Bibcode:2009Natur.462Q..11.. doi:10.1038/462011a. PMID 19890280.

[15] "UD Redd: Can the program save our tropical forests?". Thomaswhite.com. 11 May 2011. Retrieved 1 May 2013.

[16] Department of the Environment and Heritage (DEH) 2006, National Greenhouse Gas Inventory 2004: Accounting for the 108% Target, Commonwealth of Australia, Canberra.

[17] IPCC. Good Practice Guidance for Land Use, Land Use Change and Forestry. IPCC. Hayama, Japan 2003.

[18] Hohne N, Wartmann S, Herold A, Freibauer A (2007). "The rules for land use, land use change and forestry under the Kyoto Protocol—lessons learned for the future climate negotiations". *Environmental Science and Policy* **10**: 353–69. doi:10.1016/j.envsci.2007.02.001. at p. 354

[19] Department of the Environment and Heritage (DEH) 2006, National Greenhouse Gas Inventory: Analysis of Recent Trends and Greenhouse Indicators 1990 to 2004, Commonwealth of Australia, Canberra.

[20] Macintosh, Andrew (January 2007). "The National Greenhouse Accounts and Land Clearing: Do the numbers stack up?". Australia Institute. pp. 19–20. Research Paper No. 38.

[21] Earth Observatory. NASA Tropical Deforestation Research http://earthobservatory.nasa.gov/Features/Deforestation/ deforestation_update4.php accessed 12 November 2009.

[22] Spreitzer RJ, Salvucci ME (2002). "Rubisco: structure, regulatory interactions, and possibilities for a better enzyme". *Annu Rev Plant Biol* **53**: 449–75. doi:10.1146/annurev.arplant.53.100301.135233. PMID 12221984.

[23] Bond WJ, Woodward FI, Midgley GF (2005). "The global distribution of ecosystems in a world without fire". *New Phytologist* **165** (2): 525–38. doi:10.1111/j.1469-8137.2004.01252.x. PMID 15720663.

[24] Osborne, C. P.; Beerling, D. J. (2006). "Nature's green revolution: the remarkable evolutionary rise of C4 plants". *Philosophical Transactions of the Royal Society B: Biological Sciences* **361** (1465): 173–94. doi:10.1098/rstb.2005.1737. PMC 1626541. PMID 16553316.

[25] Leegood RC. (2002). "C4 photosynthesis: principles of CO_2 concentration and prospects for its introduction into C3 plants". *J. Exp. Bot.* **53** (369): 581–90. doi:10.1093/jexbot/53.369.581. PMID 11886878.

[26] Mitsue Miyao (2003). "Molecular evolution and genetic engineering of C4 photosynthetic enzymes". *J. Exp. Bot.* **54** (381): 179–89. doi:10.1093/jxb/54.381.179. PMID 12493846.

[27] Beerling, David (2008). *The Emerald Planet: How Plants Changed Earth's History*. Oxford University Press. pp. 194–5. ISBN 0-19-954814-5.

[28] Laird, David A. (2008). "The Charcoal Vision: A Win–Win–Win Scenario for Simultaneously Producing Bioenergy, Permanently Sequestering Carbon, while Improving Soil and Water Quality". *Agronomy J* **100**: 178–81. doi:10.2134/agrojnl2007.0161.

[29] Glaser B, Lehmann J, Zech W (2002). "Ameliorating physical and chemical properties of highly weathered soils in the tropics with charcoal – a review". *Biology and Fertility Soils* **35**: 219. doi:10.1007/s00374-002-0466-4.

[30] Goodall 2008, pp. 210–31

[31] Fowles M (2007). "Black carbon sequestration as an alternative to bio-energy". *Biomass and Bioenergy* **31**: 426–32. doi:10.1016/j.biombioe.2007.01.012.

[32] Andrews, Peter (2008). *Beyond the brink: Peter Andrews' radical vision for a sustainable Australian landscape*. Sydney: ABC Books for the Australian Broadcasting Corporation. p. 40. ISBN 0-7333-2410-X.

[33] Fynn, A.J., P. Alvarez, J.R. Brown, M.R. George, C. Kustin, E.A. Laca, J.T. Oldfield, T. Schohr, C.L. Neely, and C.P. Wong. 2009. "Soil carbon sequestration in U.S. rangelands" Issues paper for protocol development. Environmental Defense Fund, New York, NY, USA.

[34] Follett, R.F., Kimble, J.M., Lal, R., 2001. "The Potential of U.S. Grazing Lands to Sequester Carbon and Mitigate the Greenhouse Effect" CRC Press LLC. 1-457.

[35] "Allan Savory: How to green the desert and reverse climate change." TED Talk, February 2013.

[36] Thackara, John (June 2010). "Greener Pastures". *Seed Magazine*.

[37] Savory, Allan; Jody Butterfield (1998-12-01) [1988]. Holistic Management: A New Framework for Decision Making (2nd ed. ed.). Washington, D.C.: Island Press. ISBN 1-55963-487-1.

[38] Teague, W.R., Dowhower, S.L., Baker, S.A., Haile, N., DeLaune, P.B., Conover, D.M., (2011). "Grazing Management Impacts on Vegetation, Soil Biota and Soil Chemical, Physical and Hydrological Properties in Tall Grass Prairie" Agriculture, Ecosystems and Environment. 141. 310– 322.

[39] K.T. Weber, B.S. Gokhale, (2011). "Effect of grazing on soil-water content in semiarid rangelands of southeast Idaho" Journal of Arid Environments. 75, 464-470.

[40] Sanjari G, Ghadiri H, Ciesiolka CAA, Yu B (2008). "Comparing the effects of continuous and time-controlled grazing systems on soil characteristics in Southeast Queensland" Soil Research 46, 348–358.

[41] Retallack, Gregory (2001). "Cenozoic Expansion of Grasslands and Climatic Cooling" (PDF). *The Journal of Geology* (University of Chicago Press) **109**: 407–426. Bibcode:2001JG....109..407R. doi:10.1086/320791.

[42] Biotechnology Industry Organization (2007). *Industrial Biotechnology Is Revolutionizing the Production of Ethanol Transportation Fuel* pp. 3-4.

[43] Dale B, Kim S (2004). "Cumulative Energy and Global Warming Impact from the Production of Biomass for Biobased Products". *Journal of Industrial Ecology* **7** (3-4): 147–62. doi:10.1162/108819803323059442.

[44] Samson, R. et al. (2008). "Developing Energy Crops for Thermal Applications: Optimizing Fuel Quality, Energy Security and GHG Mitigation". In Pimentel, David. *Biofuels, Solar and Wind as Renewable Energy Systems: Benefits and*

Risks. Berlin: Springer. pp. 395–423. ISBN 1-4020-8653-9.

[45] Parr JF, Sullivan LA (2005). "Soil carbon sequestration in phytoliths". *Soil Biology and Biochemistry* **37**: 117–24. doi:10.1016/j.soilbio.2004.06.013.

[46] Tom Fearon. Australia's 'massive advantage' in biosequestration. Environmental Management News. Monday, 3 August 2009

[47] Guy Healey. Pond life fuels bio research The Australian. July 23, 2008

[48] International Energy Agency (2006). *World Energy Outlook 2006* p. 8.

[49] Garnaut 2008, p. 558

[50] Garnaut 2008, p. 432

[51] Ross Garnaut. The Garnaut Climate Change Review. Cambridge University Press, Cambridge and Melbourne 2008 ISBN 978-0-521-74444-7. p432

[52] United Nations Collaborative Programme on Reducing Emissions from Deforestation and Forest Degradation in Developing Countries *Official UN-REDD Programme Website.

[53] Stern, Nicholas Herbert (2007). *The economics of climate change: the Stern review.* Cambridge, UK: Cambridge University Press. p. xxv. ISBN 0-521-70080-9.

[54] James Hansen. Tell Barack Obama the Truth. accessed 1o Dec 2009.

[55] "Global Status of BECCS Projects 2010". Biorecro AB, Global CCS Institute. 2010. Retrieved 2012-01-20.

[56] MBD Energy Ltd. MBD captures Loy Yang Carbon Emissions. Eco Investor June 2009 http://www.mbdenergy.com/catalogue/c17/c33/p129 accessed 28 Jan 2010.

[57] Commercial scale demonstration of bio sequestration of carbon dioxide. Baird Maritime. Wednesday, 25 November 2009. http://www.bairdmaritime.com/index.php?option=com_content&view=article&id=4389:commercial-scale-demonstration-of-bio-sequestration-of-carbon-dioxide&catid=116:environment&Itemid=211 accessed 28 Jan 2010

[58] Schumacher, E. F. (1974). *Small is Beautiful: a study of economics as if people mattered.* London: Abacus. p. 112. ISBN 0-349-13139-2.

[59] Davies, Geoffrey F. (2004). *Economia: new economic systems to empower people and support the living world.* Sydney: ABC Books for the Australian Broadcasting Corporation. pp. 202–3. ISBN 0-7333-1298-5.

[60] Garnaut 2008, p. 582

[61] Saddler H and King H. Agriculture and Emissions Trading: The Impossible Dream. Australia Institute Discussion Paper 102. Australia Institute, Canberra. 2008.

[62] Collett M (2009). "In the REDD: A conservative approach to reducing emissions from deforestation and forest degradation". *CCLR* **3**: 324–39.

[63] Lohmann L. The Carbon shop: Planting New Problems. Briefing paper, Plantations Campaign, World Rainforest Movement, Moreton-in-March (UK) and Montevideo (Uruguay). 1999. p3.

[64] Humphreys 2008, p. 439

[65] Humphreys 2008, p. 440

[66] William J. Sutherland WJ et al. A horizon scan of global conservation issues for 2010 Trends in Ecology & Evolution Volume 25, Issue 1, January 2010, Pages 1-7 doi:10.1016/j.tree.2009.10.003

[67] Oscar Venter, William F. Laurance, Takuya Iwamura, Kerrie A. Wilson, Richard A. Fuller, and Hugh P. Possingham. Harnessing Carbon Payments to Protect Biodiversity. Science. 4 December. 326: 1368 (2009) doi:10.1126/science.1180289

- Garnaut, Ross (2008). *The Garnaut Climate Change Review.* Cambridge, UK: Cambridge University Press. ISBN 0-521-74444-X.

- Goodall, Chris (2008). *Ten Technologies to Save the Planet.* London: Green Profile. ISBN 1-84668-868-X.

- Humphreys, D (2008). "The politics of 'Avoided Deforestation': historical context and contemporary issues". *International Forestry Review* **10** (3): 433–42. doi:10.1505/ifor.10.3.433.

9.11   External links

- Greenfleet (not-for-profit company assisting with biosequestration options) http://www.greenfleet.com.au
- Pew Center on Global Climate Change. Biosequestration fact sheet. http://www.c2es.org/technology/factsheet/Biosequestration

- *Fungi pull carbon into northern forest soils; Organisms living on tree roots do lion's share of sequestering carbon* March 28, 2013 Vol.183 #9 Science News

Chapter 10

Community forests in England

England's twelve **community forests** are afforestation-based regeneration projects[*][1] which were established in the early 1990s.[*][2] Each of them is a partnership between the Forestry Commission and the Countryside Agency, which are agencies of the British government, and the relevant local councils.

Most of the designated areas are close to large cities and contain large amounts of brownfield, underused and derelict land. When the forests were created the average forest cover in the designated areas was 6.9%, and the target is to increase this to 30% over about 30 years. As most of the land is in private ownership the schemes rely mainly on providing landowners with incentives to plant trees. However the forests contain areas of publicly accessible open land, and increasing public access is one of the objectives.

The table below lists the twelve forests. As some of them straddle county boundaries they are more conveniently listed by region and town or city.

10.1   See also

- Forestry portal

10.2   References

[1] http://www.communityforest.org.uk/

[2] http://www.guardian.co.uk/society/2005/nov/09/
     guardiansocietysupplement1

10.3   External links

- Official site

- Forest of Avon Trust

Chapter 11

Farmer-managed natural regeneration

Farmer-managed natural regeneration (**FMNR**) is a low-cost, sustainable land-restoration technique used to combat poverty and hunger amongst poor subsistence farmers in developing countries by increasing food and timber production, and resilience to climate extremes. It involves the systematic regeneration and management of trees and shrubs from tree stumps, roots and seeds.[*][1]

FMNR is especially applicable, but not restricted to, the dryland tropics. As well as returning degraded croplands and grazing lands to productivity, it can be used to restore degraded forests, thereby reversing biodiversity loss and reducing vulnerability to climate change. FMNR can also play an important role in maintaining not-yet-degraded landscapes in a productive state, especially when combined with other sustainable land management practices such as conservation agriculture on cropland and holistic management on rangelands.[*][2]

FMNR adapts centuries-old methods of woodland management, called coppicing and pollarding, to produce continuous tree-growth for fuel, building materials, food and fodder without the need for frequent and costly replanting. On farmland, selected trees are trimmed and pruned to maximise growth while promoting optimal growing conditions for annual crops (such as access to water and sunlight).[*][3] When FMNR trees are integrated into crops and grazing pastures there is an increase in crop yields, soil fertility and organic matter, soil moisture and leaf fodder. There is also a decrease in wind and heat damage, and soil erosion.[*][4]

In the Sahel region of Africa, FMNR has become a potent tool in increasing food security, resilience and climate change adaptation in poor, subsistence farming communities where much of sub-Saharan Africa's poverty exists. FMNR is also being promoted in East Timor, Indonesia and Myanmar.

FMNR complements the evergreen agriculture,[*][5] conservation agriculture and agroforestry movements. It is considered a good entry point for resource-poor and risk-averse farmers to adopt a low-cost and low-risk technique. This in turn has acted as a stepping stone to greater agricultural intensification as farmers become more receptive to new ideas.[*][6][*][7]

11.1 Background

Throughout the developing world, immense tracts of farmland, grazing lands and forests have become degraded to the point they are no longer productive. Deforestation continues at an alarming rate. In Africa's drier regions, 74 percent of rangelands and 61 percent of rain-fed croplands are damaged by moderate to very severe desertification. In some African countries deforestation rates exceed planting rates by 300:1.[*][8]

Degraded land has an extremely detrimental effect on the lives of subsistence farmers who depend on it for their food and livelihoods. Subsistence farmers often make up to 70-80 percent of the population in these regions and they regularly suffer from hunger, malnutrition and even famine as a consequence.[*][9][*][10][*][11]

In the Sahel region of Africa, a band of savannah which runs across the continent immediately south of the Sahara Desert, large tracts of once-productive farmland are turning to desert.[*][12] In tropical regions across the world, where rich soils and good rainfall would normally assure bountiful harvests and fat livestock, some environments have become so degraded they are no longer productive. Severe famines across the African Sahel in the 1970s and 80s led to a global response, and stopping desertification became a top priority. Conventional methods of raising exotic and indigenous tree species in nurseries were used – planting out, watering, protecting and weeding. However, despite investing millions of dollars and thousands of hours labour, there was little overall impact.[*][13] Conventional approaches to reforestation in such harsh environments faced insurmountable problems and were costly and labour-intensive. Once planted out, drought, sand storms, pests, competition from weeds and destruction by people and animals negated efforts. Low levels of community ownership were another inhibiting factor.*[14]

Existing indigenous vegetation was generally dismissed as 'useless bush', and it was often cleared to make way for exotic species. Exotics were planted in fields containing living and sprouting stumps of indigenous vegetation, the presence of which was barely acknowledged, let alone seen as important.*[15]

This was an enormous oversight. In fact, these living tree stumps are so numerous they constitute a vast 'underground forest' just waiting for some care to grow and provide multiple benefits at little or no cost –and each stump can produce between 10 and 30 stems each. During the process of traditional land preparation, farmers saw the stems as weeds and slashed and burnt them before sowing their food crops. The net result was a barren landscape for much of the year with few mature trees remaining. To the casual observer, the land was turning to desert. Most concluded that there were no trees present and that the only way to reverse the problem was through tree planting.*[16]

Meanwhile, established indigenous trees continued to disappear at an alarming rate. In Niger, from the 1930s until 1993, forestry laws took tree ownership and responsibility for the care of trees out of the hands of the people; and even though ineffective and uneconomic, reforestation through conventional tree planting seemed to be the only way to address desertification at the time.*[17]*[18]*[19]

11.2 Birth and spread

In the early 1980s, in the Maradi region of the Republic of Niger, the missionary organisation, Serving in Mission (SIM), was unsuccessfully attempting to reforest the surrounding districts using conventional means. In 1983, SIM began experimenting and promoting FMNR amongst about 10 farmers. During the severe famine of 1984, a food-for-work program was introduced that saw some 70,000 people exposed to FMNR and its practice on around 12,500 hectares of farmland. From 1985 to 1999, FMNR continued to be promoted locally and nationally as exchange visits and training days were organised for various NGOs, government foresters, Peace Corps Volunteers and farmer and civil society groups. Additionally, SIM project staff and farmers visited numerous locations across Niger to provide training.*[20]

By 2004 it was ascertained that FMNR was being practiced on over five million hectares or 50 percent of Niger's farmland – an average reforestation rate of 250,000 hectares per year over a 20-year period. This transformation prompted Senior Fellow of the World Resources Institute, Chris Reij to comment that "this is probably the largest positive environmental transformation in the Sahel and perhaps all of Africa".*[21]*[22]

Also in 2004, World Vision Australia and World Vision Ethiopia initiated a forestry-based carbon sequestration project as a potential means to stimulate community development while engaging in environmental restoration. An innovative partnership with the World Bank, the Humbo Community-based Natural Regeneration Project involved the regeneration of 2,728 hectares of degraded native forests. This brought social, economic and ecological benefits to the participating communities. Within two years, communities were collecting wild fruits, firewood and fodder, and reported that wildlife had begun to return and erosion and flooding had been reduced. In addition, the communities are now receiving payments for the sale of carbon credits through the Clean Development Mechanism (CDM) of the Kyoto protocol.*[23]*[24]

Following the success of the Humbo project, FMNR spread to the Tigray region of northern Ethiopia where 20,000 hectares have been set aside for regeneration, including 10-hectare FMNR model sites for research and demonstration in each of 34 sub-districts.*[25] In addition, the Government of Ethiopia has committed to reforest 15 million hectares of degraded land using FMNR as part of a climate change and renewable energy plan to become carbon neutral by 2025.*[26]

In Talensi, northern Ghana, FMNR is being practiced on 2,000-3,000 hectares and new projects, initiated by World Vision, are introducing FMNR into three new districts. In the Kaffrine and Diourbel regions of Senegal, FMNR has spread across 50,000 hectares in four years.*[27] World Vision is also promoting FMNR in Indonesia, Myanmar and East Timor.*[28] There are also examples of both independently promoted and spontaneous FMNR movements occurring. In Burkina Faso, for example, an increasing part of the country is being transformed into agroforestry parkland. And in Mali, an ageing agroforestry parkland of about 6 million hectares is showing signs of regeneration.*[29]*[30]

11.3 Key principles

FMNR depends on the existence of living tree stumps or roots in crop fields, grazing pastures, woodlands or forests. Each season bushy growth will sprout from the stumps/roots often appearing like small shrubs. Continuous grazing by livestock, regular burning and/or regular harvesting for fuel wood results in these 'shrubs' never attaining tree stature. On farmland, standard practice has been for farmers to slash this regrowth in preparation for planting crops, but with a little attention this growth can be turned into a valuable resource without jeopardising, but in fact, enhancing crop yields.*[31]*[32]

For each stump, a decision is made as to how many stems will be chosen to grow. The tallest and straightest stems are selected and the remaining stems culled. Best results are obtained when the farmer returns regularly to prune any unwanted new stems and side branches as they appear. Farmers can then grow other crops between and around the trees. When farmers want wood they can cut the stem(s) they want and leave the rest to continue growing. The remaining stems will increase in size and value each year, and will continue to protect the environment. Each time a stem is harvested, a younger stem is selected to replace it.*[33]

Various naturally occurring tree species can be used which may also provide berries, fruits and nuts or have medicinal qualities. In Niger, commonly used species include: Strychnos spinosa, Balanites aegyptiaca, Boscia senegalensis, Ziziphus spp., Annona senegalensis, Poupartia birrea and Faidherbia albida. However, the most important determinants are whatever species are locally available, their ability to re-sprout after cutting, and the value local people place on those species.*[34]

Faidherbia albida, also known as the 'fertiliser tree', is popular for intercropping across the Sahel as it fixes nitrogen into the soil, provides fodder for livestock, and shade for crops and livestock. By shedding its leaves in the wet season, Faidherbia provides beneficial light shade to crops when high temperatures would otherwise damage crops or retard growth. Leaf fall contributes useful nutrients and organic matter to the soil.*[35]

The practice of FMNR is not confined to croplands. It is being practised on grazing land and in degraded communal forests as well. When there are no living stumps, seeds of naturally occurring species are used. In reality, there is no fixed way of practising FMNR and farmers are free to choose which species they will leave, the density of trees they prefer, and the timing and method of pruning.*[36]

11.4   FMNR in practice

FMNR depends on the existence of living tree stumps, tree roots and seeds to be re-vegetated. These can be in crop fields, grazing lands or degraded forests. New stems, which sprout from these stumps and tree roots, can be selected and pruned for improved growth. Sprouting tree stumps and roots may look like shrubs and are often ignored or even slashed by farmers or foresters. However, with culling of excess stems and by selecting and pruning of the best stems, the re-growth has enormous potential to rapidly grow into trees.

Seemingly treeless fields may contain seeds and living tree stumps and roots which have the ability to sprout new stems and regenerate trees. Even this 'bare' millet field in West

Africa contains hundreds of living stumps per hectare which are buried beneath the surface like an underground forest.

Step 1. Do not automatically slash all tree growth, but survey your farm noting how many and what species of trees are present.

Step 2. Select the stumps which will be used for regeneration.

Step 3. Select the best five or so stems and cull unwanted ones. This way, when you want wood you can cut the stem(s) that are needed and leave the rest to continue growing. These remaining stems will increase in size and value each year, and will continue to protect the environment and provide other useful materials and services such as fodder, humus, habitat for useful pest predators and protection from the wind and sun. Each time one stem is harvested, a younger stem is selected to replace it.

Tag selected stems with a coloured rag or paint. Work with the whole community to draw up and agree on laws which will protect the trees being pruned and respect each person's rights. Where possible, include government forestry staff and local authorities in planning and decision making.

11.5 Benefits of FMNR

FMNR can restore degraded farmlands, pastures and forests by increasing the quantity and value of woody vegetation, by increasing biodiversity and by improving soil structure and fertility through leaf litter and nutrient cy-

cling. The reforestation also retards wind and water erosion; it creates wind-breaks which decrease soil moisture evaporation, and protects crops and livestock against searing winds and temperatures. Often, dried up springs reappear and the water table rises towards historic levels; insect eating predators including insects, spiders and birds return, helping to keep crop pests in check; the trees can be a source of edible berries and nuts; and over time the biodiversity of plant and animal life is increased.*[37]*[38] FMNR can be used to combat deforestation and desertification and can also be an important tool in maintaining the integrity and productivity of land that is not yet degraded.

Trials, long-running programs and anecdotal data indicate that FMNR can at least double crop yields on low fertility soils.*[39] In the Sahel, high numbers of livestock and an eight-month dry season can mean that pastures are completely depleted before the rains commence. However, with the presence of trees, grazing animals can make it through the dry season by feeding on tree leaves and seed pods of some species, at a time when no other fodder is available.*[40]*[41] In north east Ghana, more grass became available with the introduction of FMNR because communities worked together to prevent bush fires from destroying their trees.*[42]

Well designed and executed FMNR projects can act as catalysts to empower communities as they negotiate land ownership or user rights for the trees in their care. This assists with self-organisation, and with the development of new agriculture-based micro-enterprises (e.g. selling firewood, timber and handcrafts made from timber or woven grasses).*[43]

Conventional approaches to reversing desertification, such as funding tree planting, rarely spread beyond the project boundary once external funding is withdrawn. By comparison, FMNR is cheap, rapid, locally led and implemented. It uses local skills and resources – the poorest farmers can learn by observation and teach their neighbours. Given an enabling environment, or at least the absence of a 'disabling' environment, FMNR can be done at scale and spread well beyond the original target area without ongoing government or NGO intervention.[44]

World Vision evaluations of FMNR conducted in Senegal and Ghana in 2011 and 2012 found that households practising FMNR were less vulnerable to extreme weather shocks such as drought and damaging rain and wind storms.[45][46]

The following table summarises FMNR's benefits which fit the sustainable development model of economic, social and environmental benefits:

Source: compiled by R Francis, Project Manager FMNR, World Vision Australia from Brown et al,[47] Garrity et al,[48] Haglund et al,[49] McGahuey & Winterbottom,[50] Reij et al,[51] Rinaudo,[52][53] World Recources Institute.[54]

11.6 Key success factors and constraints

While there are numerous accounts of the uptake and spread of FMNR independent of aid and development agencies, the following factors have been found to be beneficial for its introduction and spread:

- Awareness creation of FMNR's potential.

- Capacity building through workshops and exchange visits.

- Awareness of the devastating effects of deforestation. The adoption of FMNR is more likely when communities acknowledge their situation and the need to take action. This perception of need can be supported by education.

- An FMNR champion/facilitator from within the community who encourages, challenges and trains peers. This is critical during the first three to five years, and continues to be important for up to 10 years. Regular site visits also ensure early detection and remedial action on resistance and threats to FMNR through deliberate damage to trees and theft.

- The buy-in of all stakeholders including their agreement on any by-laws created for FMNR and the consequences for infringements. Stakeholders include FMNR practitioners, local, regional and national government departments of agriculture and forestry, men, women, youth, marginalised groups (including nomadic herders), cultivators and commercial interests.

- Stakeholder buy-in is also important to create a critical mass of FMNR adopters in order to change social attitudes from a position of apathy or active participation in deforestation to one of proactive sustainable tree management through FMNR.

- Government support through the creation of favourable policies, positive reinforcement of actions facilitating the spread of FMNR, and disincentives for actions working against the spread of FMNR. FMNR practitioners need to be confident that they will benefit from their labours (either private or community ownership of trees, or legally binding user rights).

- Reinforcement of existing organisational structures (farmers clubs, development groups, traditional leadership structures) or establishment of new structures which will provide a framework for communities to practise FMNR on a local, district or region-wide basis.

- A communications strategy which includes education in schools, radio programs and engagement with religious and traditional leaders to become advocates.

- Establishment of a legal, transparent and accessible market for FMNR wood and non-timber forest products, enabling practitioners to benefit financially from their activities.

Brown et al. suggest that the two main reasons why FMNR has spread so widely in Niger are attitudinal change by the community of what constitutes good land management practices, and farmers' ownership of trees.[55] Farmers need the assurance that they will benefit from their labour. Giving farmers either outright ownership of the trees they protect, or tree-user rights, has made it possible for large-scale farmer-led reforestation to take place.[56]

11.7 Current and future directions

Over nearly 30 years, FMNR has changed the farming landscape in some of the poorest countries in the world, including parts of Niger, Burkina Faso, Mali and Senegal, providing subsistence farmers with the methods necessary to become more food secure and resilient against severe weather events.[57][58]

The 2011–12 food crisis in East Africa gave a stark reminder of the importance of addressing root causes of hunger. In the 2011 State of the World Report, Bunch concludes that four major factors – lack of sustainable fertile land, loss of traditional fallowing, cost of fertiliser and climate change – are coming together all at once in a sort of "perfect storm" that will almost surely result in an African famine of unprecedented proportions, probably within the next four to five years. It will most heavily affect the lowland, semi-arid to sub-humid areas of Africa (including the Sahel, parts of eastern Africa, plus a band from Malawi across to Angola and Namibia); and unless the world does something dramatic, 10 to 30 million people could die from famine between 2015 and 2020.[59] Restoration of degraded land through FMNR is one way of addressing these major contributors to hunger.

In recent years FMNR has come to the attention of global development agencies and grass roots movements alike. The World Bank, World Resources Institute, World Agroforestry Center, USAID and the Permaculture movement are amongst those either actively promoting or advocating for the uptake of FMNR and FMNR has received recognition from a number of quarters including:

- In 2010, FMNR won the Interaction 4 Best Practice and Innovation Initiative award in recognition of high technical standards and effectiveness in addressing the food security and livelihood needs of small producers in the areas of natural resource management and agro forestry.

- In 2011, FMNR won the World Vision International Global Resilience Award for the most innovative initiative in the area of resilient development practice and natural environment and climate issues.

- In 2012 FMNR won the Arbor Day Award for Education Innovation.[60]

In April 2012, World Vision Australia – in partnership with the World Agroforestry Center and World Vision East Africa – held an international conference in Nairobi called Beating Famine to analyse and plan how to improve food security for the world's poor through the use of FMNR and Evergreen Agriculture. The conference was attended by more than 200 participants, including world leaders in sustainable agriculture, five East African ministers of agriculture and the environment, ambassadors and other government representatives from Africa, Europe and Australia, and leaders from non-government and international organisations.

Two major outcomes of the conference were:

1. The establishment of a global FMNR network of key stakeholders to promote, encourage and initiate the scale-up of FMNR globally.

2. Country, regional and global level plans as a basis for inter-organisation collaboration for FMNR scale-up.

The conference acted as a catalyst for media coverage of FMNR in some of the world's leading outlets and a noticeable increase in momentum for an FMNR global movement. This heightened awareness of FMNR has created an opportunity for it to spread exponentially worldwide.[61][62]

World Vision and the World Agroforestry Centrer are currently exploring opportunities for conducting conferences and workshops in new regions where FMNR is not yet established in order to stimulate further awareness and adoption.

11.8   See also

- Agroforestry

- Sustainable forest management

- Regenerative Agriculture

- Zaï

11.9   References

[1] Rinaudo, T 2012, Natural Resources Advisor, World Vision Australia and pioneer of Farmer Managed Natural Regeneration in Niger in 1983; and Francis, R 2012, Project Manager, Farmer Managed Natural Regeneration, World Vision Australia.

[2] Rinaudo, T 2012, Natural Resources Advisor, World Vision Australia and pioneer of Farmer Managed Natural Regeneration in Niger in 1983.

[3] World Resources Institute (WRI) in collaboration with United Nations Development Programme, United Nations Environment Programme, and World Bank. 2008, Turning Back the Desert: How Farmers Have Transformed Niger's Landscapes and Livelihoods, Chapter 3 in World Resources 2008: Roots of Resilience – Growing the Wealth of the Poor, http://www.wri.org/publication/world-resources-2008-roots-of-resilience.

[4] Rinaudo, T 2012, Natural Resources Advisor, World Vision Australia and pioneer of Farmer Managed Natural Regeneration in Niger in 1983.

[5] Evergreen Agriculture – the incorporation of trees into crop land and pastures. The three types of evergreen agriculture are: Farmer Managed Natural Regeneration, the planting of trees in conventional crop fields, and conservation agriculture with trees. Evergreen agriculture is one of several types of agroforestry.

[6] Garrity, D 2012, UNCCD Drylands Ambassador, Distinguished Board Research Fellow and former Director General of the World Agroforestry Centre, Chairman of Landcare International.

[7] Rinaudo, T 2012, Natural Resources Advisor, World Vision Australia and pioneer of Farmer Managed Natural Regeneration in Niger in 1983.

[8] Grainger, A 1990, The Threatening Desert: Controlling desertification, pp 3, 141, 142, 215, 298, Earthscan, London.

[9] Brown, D, Dettmann, P, Rinaudo, T, Tefera, H and Tofu, A 2010, Poverty Alleviation and Environmental Restoration Using the Clean Development Mechanism: A Case Study from Humbo, Ethiopia, Environmental Management (2011) 48:322–333.

[10] Garrity, D, Akinnifesi, F, Ajayi, O, Weldesemayat, S, Mowo, J, Kalinganire, A, Larwanou, M and Bayala, J 2010, Evergreen Agriculture: a robust approach to sustainable food security in Africa, Food Security (2010) 2:197–214.

[11] Rinaudo, T 2007, The Development of Farmer Managed Natural Regeneration, LEISA Magazine 23.2 June 2007, Ileia Centre for Learning on Sustainable Agriculture, Wageningen, Netherlands.

[12] Tougiani, A, Guero, C and Rinaudo, T 2008, Community mobilisation for improved livelihoods through tree crop management in Niger, GeoJournal (2009) 74:377–389.

[13] Grainger, A 1990, The Threatening Desert: Controlling desertification, p 301, Earthscan, London.

[14] Rinaudo, T 2007, The Development of Farmer Managed Natural Regeneration, LEISA Magazine 23.2 June 2007, Ileia Centre for Learning on Sustainable Agriculture, Wageningen, Netherlands.

[15] Rinaudo, T 2012, Natural Resources Advisor, World Vision Australia and pioneer of Farmer Managed Natural Regeneration in Niger in 1983.

[16] Rinaudo, T 2007, The Development of Farmer Managed Natural Regeneration, LEISA Magazine 23.2 June 2007, Ileia Centre for Learning on Sustainable Agriculture, Wageningen, Netherlands.

[17] Brough, W and Kimenyi, M 2004, "Desertification" of the Sahel: Exploring the role of property rights, Property and Environment Research Center, Vol 2, No. 2, June 2004, http://perc.org/articles/desertification-sahel.

[18] World Resources Institute (WRI) in collaboration with United Nations Development Programme, United Nations Environment Programme, and World Bank. 2008, Turning Back the Desert: How Farmers Have Transformed Niger's Landscapes and Livelihoods, Chapter 3 in World Resources 2008: Roots of Resilience – Growing the Wealth of the Poor, http://www.wri.org/publication/world-resources-2008-roots-of-resilience.

[19] Rinaudo, T 2007, The Development of Farmer Managed Natural Regeneration, LEISA Magazine 23.2 June 2007, Ileia Centre for Learning on Sustainable Agriculture, Wageningen, Netherlands.

[20] Rinaudo, T 2007, The Development of Farmer Managed Natural Regeneration, LEISA Magazine 23.2 June 2007, Ileia Centre for Learning on Sustainable Agriculture, Wageningen, Netherlands.

[21] World Resources Institute (WRI) in collaboration with United Nations Development Programme, United Nations Environment Programme, and World Bank. 2008, Turning Back the Desert: How Farmers Have Transformed Niger's Landscapes and Livelihoods, Chapter 3 in World Resources 2008: Roots of Resilience – Growing the Wealth of the Poor, http://www.wri.org/publication/world-resources-2008-roots-of-resilience.

[22] Hertsgaard, M 2009, Regreening Africa, The Nation, 7 December 2009, http://www.thenation.com/article/regreening-africa.

[23] Brown, D, Dettmann, P, Rinaudo, T, Tefera, H and Tofu, A 2010, Poverty Alleviation and Environmental Restoration Using the Clean Development Mechanism: A Case Study from Humbo, Ethiopia, Environmental Management (2011) 48:322–333.

[24] The World Bank 2012, Ethiopia: Humbo Assisted Natural Regeneration, http://wbcarbonfinance.org/Router.cfm?Page=Projport&ProjID=9625.

[25] Rinaudo, T 2012, Natural Resources Advisor, World Vision Australia and pioneer of Farmer Managed Natural Regeneration in Niger in 1983.

[26] UNDP 2012, Programme document for Ethiopia (2012–2015), www.et.undp.org/index.php?option=com_docman&task...

[27] World Vision Australia Monitoring and Evaluation Reports (internal documents).

[28] Rinaudo, T 2012, Natural Resources Advisor, World Vision Australia and pioneer of Farmer Managed Natural Regeneration in Niger in 1983.

[29] World Agroforestry Center 2012, Improving Sustainable Productivity in Farming Systems and Enhanced Livelihoods through Adoption of Evergreen Agriculture in Eastern Africa, Research Proposal to the Australian Centre for International Agricultural Research, World Agroforestry Center, Nairobi.

[30] Reij, C 2012, African Re-greening Initiatives Blog, see http://www.africa-regreening.blogspot.com.au.

[31] Ndour, B, Sarr, A et Mbaye, A 2010, Projects Baysatol/SFLEI – Rapport d'Activities, Institut Sénégalais de Recherches Agricoles (ISRA), Centre National de Recherches Agronomiques, Replublique du Senegal Ministere de l'Agriculture / World Vision Senegal.

[32] Garrity, D, Akinnifesi, F, Ajayi, O, Weldesemayat, S, Mowo, J, Kalinganire, A, Larwanou, M and Bayala, J 2010, Evergreen Agriculture: a robust approach to sustainable food security in Africa, Food Security (2010) 2:197–214.

[33] Rinaudo, T 2012, Natural Resources Advisor, World Vision Australia and pioneer of Farmer Managed Natural Regeneration in Niger in 1983.

[34] Rinaudo, T 2012, Natural Resources Advisor, World Vision Australia and pioneer of Farmer Managed Natural Regeneration in Niger in 1983.

[35] Garrity, D, Akinnifesi, F, Ajayi, O, Weldesemayat, S, Mowo, J, Kalinganire, A, Larwanou, M and Bayala, J 2010, Evergreen Agriculture: a robust approach to sustainable food security in Africa, Food Security (2010) 2:197–214.

[36] Rinaudo, T 2012, Natural Resources Advisor, World Vision Australia and pioneer of Farmer Managed Natural Regeneration in Niger in 1983.

[37] Rinaudo, T 2007, The Development of Farmer Managed Natural Regeneration, LEISA Magazine 23.2 June 2007, Ileia Centre for Learning on Sustainable Agriculture, Wageningen, Netherlands.

[38] World Vision Australia 2012, Turning 'Deserts' into Food Bowls by Releasing the Underground Forest – Latest developments in Farmer Managed Natural Regeneration, internal briefing paper, World Vision Australia, Melbourne.

[39] Ndour, B, Sarr, A et Mbaye, A 2010, Projects Baysatol/SFLEI – Rapport d' Activities, Institut Sénégalais de Recherches Agricoles (ISRA), Centre National de Recherches Agronomiques, Replublique du Senegal Ministere de l' Agriculture / World Vision Senegal.

[40] Reij, C, Tappan, G, and Smale, M 2009. Re-Greening the Sahel: Farmer-led innovation in Burkina Faso and Niger, Chapter 7 in: Spielman, D, Pandya-Lorch, R (eds): Millions Fed – proven successes in agricultural development, International Food Policy Research Institute, US, pp 53-58, see http://www.ifpri.org/sites/default/files/publications/oc64ch07.pdf.

[41] World Vision Australia 2012, Turning 'Deserts' into Food Bowls by Releasing the Underground Forest – Latest developments in Farmer Managed Natural Regeneration, internal briefing paper, World Vision Australia, Melbourne.

[42] Rinaudo, T 2012, Natural Resources Advisor, World Vision Australia and pioneer of Farmer Managed Natural Regeneration in Niger in 1983.

[43] World Vision Australia 2012, Turning 'Deserts' into Food Bowls by Releasing the Underground Forest – Latest developments in Farmer Managed Natural Regeneration, internal briefing paper, World Vision Australia, Melbourne.

[44] Rinaudo, T 2012, Natural Resources Advisor, World Vision Australia and pioneer of Farmer Managed Natural Regeneration in Niger in 1983.

[45] Diagne, M 2012, Evaluation Finale du Projet Beysatol, World Vision Senegal.

[46] Weston, P and Hong, R 2012, End-of-Phase Evaluation Report: Talensi FMNR Project, World Vision Australia / World Vision Ghana.

[47] Brown, D, Dettmann, P, Rinaudo, T, Tefera, H and Tofu, A 2010, Poverty Alleviation and Environmental Restoration Using the Clean Development Mechanism: A Case Study from Humbo, Ethiopia, Environmental Management (2011) 48:322–333.

[48] Garrity, D, Akinnifesi, F, Ajayi, O, Weldesemayat, S, Mowo, J, Kalinganire, A, Larwanou, M and Bayala, J 2010, Evergreen Agriculture: a robust approach to sustainable food security in Africa, Food Security (2010) 2:197–214.

[49] Haglund, E, Ndjeunga, J, Snook, L and Pasternak, D 2009, Assessing the Impacts of Farmer Managed Natural Regeneration in the Sahel: A Case Study of Maradi Region, Niger, internal report for International Crops Research Institute for the Semi-Arid Tropics, Niamey, Niger.

[50] McGahuey, M and Winterbottom R, 2007, Transformational Development in Niger, Presentation to US Aid, see US Aid Frame Natural Resource Management Communities, http://www.frameweb.org/CommunityBrowser.aspx?id=2803&lang=en-US.

[51] Reij, C., Tappan, G., and Smale, M. 2009. Re-Greening the Sahel: Farmer-led innovation in Burkina Faso and Niger, Chapter 7 in: Spielman, D., Pandya-Lorch, R (eds): Millions Fed – proven successes in agricultural development, International Food Policy Research Institute, US, pp.53-58 Available online: http://www.ifpri.org/sites/default/files/publications/oc64ch07.pdf.

[52] Rinaudo, T 2005, Economic Benefits of Farmer Managed Natural Regeneration, World Vision Australia Briefing Note, see US Aid Frame Natural Resource Management Communities, http://www.frameweb.org/CommunityBrowser.aspx?id=2840&lang=en-US.

[53] Rinaudo, T 2012, Natural Resources Advisor, World Vision Australia and pioneer of Farmer Managed Natural Regeneration in Niger in 1983.

[54] World Resources Institute (WRI) in collaboration with United Nations Development Programme, United Nations Environment Programme, and World Bank. 2008, Turning Back the Desert: How Farmers Have Transformed Niger's Landscapes and Livelihoods, Chapter 3 in World Resources 2008: Roots of Resilience—Growing the Wealth of the Poor, Washington, see http://www.wri.org/publication/world-resources-2008-roots-of-resilience.

[55] Brown, D, Dettmann, P, Rinaudo, T, Tefera, H and Tofu, A 2010, Poverty Alleviation and Environmental Restoration Using the Clean Development Mechanism: A Case Study from Humbo, Ethiopia, Environmental Management (2011) 48:322–333.

[56] Rinaudo, T 2012, Natural Resources Advisor, World Vision Australia and pioneer of Farmer Managed Natural Regeneration in Niger in 1983.

[57] Reij, C, Tappan, G, and Smale, M 2009. Re-Greening the Sahel: Farmer-led innovation in Burkina Faso and Niger, Chapter 7 in: Spielman, D, Pandya-Lorch, R (eds): Millions Fed – proven successes in agricultural development, International Food Policy Research Institute, US, pp 53-58, see http://www.ifpri.org/sites/default/files/ publications/oc64ch07.pdf.

[58] Hertsgaard, M 2011, The Great Green Wall: African Farmers Beat Back Drought and Climate Change with Trees – A quiet, green miracle has been growing in the Sahel, excerpt from Hertsgaard, M 2011, Hot: Living Through the Next 50 Years on Earth, publ Houghton Miflin Harcourt in Scientific American 28 January 2011, see http://www.scientificamerican.com/article.cfm?id= farmers-in-sahel-beat-back-drought-and-climate-change-with-trees.

[59] Bunch, R 2011, Africa's Soil Fertility Crisis and the Coming Famine, Chapter 6 in the Worldwatch Institute report State of the World 2011: Innovations that Nourish the Planet, see http://www.worldwatch.org/sow11.

[60] 2012 Arbor Day Awards, Nebraska City, US, see http://www.arborday.org/programs/awards/2012/?award= Education%20Innovation.

[61] Francis, R 2012, Project Manager, Farmer Managed Natural Regeneration, World Vision Australia.

[62] Guardian/Observer 26 August 2012, Africa innovations: 15 ideas helping to transform a continent, http://www.guardian.co.uk/world/2012/aug/ 26/africa-innovations-transform-continent; ABC Lateline 9 July 2012, Reforestation project adds hope to food crisis, Reporter Ginny Stein, http: //www.abc.net.au/lateline/content/2012/s3542254.htm; PBS Newshour 12 July 2012, Amidst Drought and Famine, Niger Leads West Africa in Addressing Crisis, http://www. pbs.org/newshour/bb/africa/july-dec12/niger_07-12.html; Der Spiegel Online 18 June 2012, Begrünung in der Sahelzone – Mister Rinaudo will die Wüste, Aus Nairobi berichtet Horand Knaup, http://www.spiegel.de/wissenschaft/natur/ sahelzone-fmnr-soll-ausbreitung-der-wueste-stoppen-a-838840. html.

11.10   External links

- Farmer Managed Natural Regeneration Website: http: //www.fmnrhub.com.au/

- Re-Greening the Sahel at IFPRI

- The Development of Farmer Managed Natural Regeneration

- Farmer Managed Natural Regeneration - Video

Chapter 12

Forest landscape restoration

Forest landscape restoration (FLR) is defined as "a planned process to regain ecological integrity and enhance human well-being in deforested or degraded landscapes".*[1] It comprises tools and procedures to integrate site-level forest restoration actions with desirable landscape-level objectives, which are decided upon via various participatory mechanisms among stakeholders. The concept has grown out of collaboration among some of the world's major international conservation organizations including the International Union for Conservation of Nature (IUCN), the World Wide Fund for Nature (WWF), the World Resources Institute and the International Tropical Timber Organization (ITTO).

12.1 Aims

The concept of FLR was conceived to bring about compromises between meeting the needs of both humans and wildlife, by restoring a range of forest functions at the landscape level. It includes actions to strengthen the resilience and ecological integrity of landscapes and thereby keep future management options open. The participation of local communities is central to the concept, because they play a critical role in shaping the landscape and gain significant benefits from restored forest resources. Therefore, FLR activities are inclusive and participatory.*[2]

12.1.1 Desirable outcomes

The desirable outcomes of an FLR program usually comprise a combination of the following, depending on local needs and aspirations:

- identification of the root causes of forest degradation and prevention of further deforestation,

- positive engagement of people in the planning of forest restoration, resolution of land-use conflicts and agreement on benefit-sharing systems,

- compromises over land-use trade-offs that are acceptable to the majority of stakeholders,

- a repository of biological diversity of both local and global value,

- delivery of a range of utilitarian benefits to local communities including:

  - a reliable supply of clean water,

  - environmental protection particularly watershed services (e.g. reduced soil erosion, lower landslide risk, flood/drought mitigation etc.),

  - a sustainable supply of a diverse range of forest products including foods, medicines, firewood etc.,

  - monetary income from various sources e.g. ecotourism, carbon trading via the REDD+ mechanism and from payments for other environmental services (PES)*[3]

12.2 Activities

FLR combines several existing principles and techniques of development, conservation and natural resource management, such as landscape character assessment, participatory rural appraisal, adaptive management etc. within a clear and consistent evaluation and learning framework. An FLR program may comprise various forestry practices on different sites within the landscape, depending on local environmental and socioeconomic factors. These may include protection and management of secondary and degraded primary forests, standard forest restoration techniques such as "assisted" or "accelerated" natural regeneration (ANR) and the planting of framework tree species to restore degraded areas, as well as conventional tree plantations and agroforestry systems to meet more immediate monetary needs *[4]

The IUCN hosts the The Global Partnership on Forest Landscape Restoration, which co-ordinates development of the concept around the world.

12.3   See also

- Forest restoration

- Reforestation

12.4   References

[1] Reitbergen-McCraken, J., S. Maginnis A. Sarre, 2007. *The Forest Landscape Restoration Handbook*. Earthscan, London, 175 pp

[2] Lamb, D., 2011. *Regreening the Bare Hills*. Springer 547pp.

[3] Mansourian, S., D. Vallauri, and N. , Dudley (eds.) (in cooperation with WWF International), *Forest Restoration in Landscapes: Beyond Planting Trees*. Springer, New York.

[4] Elliott, S., D. Blakesley and K. Hardwick, in press. *Restoring Tropical Forests: a Practical Guide*. Kew Publications, London

12.5   External links

- Collaborative Forest Landscape Restoration Program, United States Forest Service, Washington, DC

- Forest and Landscape Restoration Project, World Resources Institute, Washington, DC

- The Global Partnership on Forest and Landscape Restoration, Wageningen, Netherlands

Chapter 13

Forest restoration

In the 1980s, conservation organizations warned that, once destroyed, tropical forests could never be restored. Thirty years of restoration research now challenge this: a) This site in Doi Suthep-Pui National Park, N. Thailand was deforested, over-cultivated and then burnt. The black tree stump was one of the original forest trees. Local people teamed up with scientists to repair their watershed.

b) Fire prevention, nurturing natural regeneration and planting framework tree species resulted in trees growing above the weed canopy within a year.

Forest restoration is defined as "actions to re-instate ecological processes, which accelerate recovery of forest structure, ecological functioning and biodiversity levels towards

c) After 12 years, the restored forest overwhelmed the black tree stump.

those typical of climax forest" *[1] i.e. the end-stage of natural forest succession. Climax forests are relatively stable ecosystems that have developed the maximum biomass, structural complexity and species diversity that are possible *within the limits imposed by climate and soil and without continued disturbance from humans* (more explanation here). Climax forest is therefore the target ecosystem, which defines the ultimate aim of forest restoration. Since climate is a major factor that determines climax forest composition, global climate change may result in changing restoration aims.*[2]

Forest restoration is a specialized form of reforestation, but it differs from conventional tree plantations in that its primary goals are biodiversity recovery and environmental protection.*[3]*[4]

13.1 Scope

Forest restoration may include simply protecting remnant vegetation (fire prevention, cattle exclusion etc.) or more active interventions to accelerate natural regeneration,*[5] as well as tree planting and/or sowing seeds (direct seeding) of species characteristic of the target ecosystem. Tree species planted (or encouraged to establish) are those that are typical of, or provide a critical ecological function in, the target ecosystem. However, wherever people live in or near restoration sites, restoration projects often include economic species amongst the planted trees, to yield subsistence or cash-generating products.

Forest restoration is an inclusive process, which depends on collaboration among a wide range of stakeholders including local communities, government officials, non-government organizations, scientists and funding agencies. Its ecological success is measured in terms of increased biological diversity, biomass, primary productivity, soil organic matter and water-holding capacity, as well as the return of rare and keystone species, characteristic of the target ecosystem. Economic indices of success include the value of forest products and ecological services generated (e.g. watershed protection, carbon storage etc.), which ultimately contribute towards poverty reduction. Payments for such ecological services (PES) and forest products can provide strong incentives for local people to implement restoration projects.

13.2   Opportunities for forest restoration

Demonstration forest restoration plot, SUNY-ESF, Syracuse, NY

Forest restoration is appropriate wherever biodiversity recovery is one of the main goals of reforestation, such as for wildlife conservation, environmental protection, ecotourism or to supply a wide variety of forest products to local communities. Forests can be restored in a wide range of circumstances, but degraded sites within protected areas are a high priority, especially where some climax forest remains

as a seed source within the landscape. Even in protected areas, there are often large deforested sites: logged over areas or sites formerly cleared for agriculture. If protected areas are to act as Earth's last wildlife refuges, restoration of such areas will be needed.*[6]*[7]

Many restoration projects are now being implemented under the umbrella of "forest landscape restoration" (FLR),*[8] defined as a "planned process to regain ecological integrity and enhance human well-being in deforested or degraded landscapes". FLR recognizes that forest restoration has social and economic functions. It aims to achieve the best possible compromise between meeting both conservation goals and the needs of rural communities.*[9] As human pressure on landscapes increases, forest restoration will most commonly be practiced within a mosaic of other forms of forest management, to meet the economic needs of local people.

13.3   Natural regeneration

Tree planting is not always essential to restore forest ecosystems. A lot can be achieved by studying how forests regenerate naturally, identifying the factors that limit regeneration and devising methods to overcome them. These can include weeding and adding fertilizer around natural tree seedlings, preventing fire, removing cattle and so on. This is "accelerated" or "assisted" natural regeneration.*[10] It is simple and cost-effective, but it can only operate on trees that are already present, mostly light-loving pioneer species. Such tree species are not usually those that comprise climax forests, but they can foster recolonization of the site by shade-tolerant climax forest tree species, via natural seed dispersal from remnant forest. Because this is a slow process, biodiversity recovery can usually be accelerated by planting some climax forest tree species, especially large-seeded, poorly dispersed species. It is not feasible to plant all the tree species that may have formerly grown in the original primary forest and it is usually unnecessary to do so, if the framework species method*[11]*[12] can be used.

13.4   See also

- Forest landscape restoration

- Reforestation

- Restoration ecology

- Secondary forest

13.5 References

[1] Elliott, S., D. Blakesley and K. Hardwick, in press. Restoring Tropical Forests: a Practical Guide. Kew Publications, London

[2] Sgró, C.M., A. J. Lowe and A. A. Hoffmann, 2011. Building evolutionary resilience for conserving biodiversity under climate change. Evolutionary Applications 4 (2): 326-337

[3] Lamb, D., 2011.Regreening the Bare Hills. Springer 547pp.

[4] http://www.treesearch.fs.fed.us/pubs/22209

[5] "Assisted natural regeneration of forests" .

[6] Lamb, D., 2011.Regreening the Bare Hills. Springer 547pp.

[7] http://www.treesearch.fs.fed.us/pubs/22209

[8] Mansourian, S., D. Vallauri, and N. Dudley (eds.) (in cooperation with WWF International), 2005. Forest Restoration in Landscapes: Beyond Planting Trees. Springer, New York

[9] Reitbergen-McCraken, J., S. Maginnis A. Sarre, 2007. The Forest Landscape Restoration Handbook. Earthscan, London, 175 pp.

[10] Shono, K., E. A. Cadaweng and P. B. Durst, 2007. Application of Assisted Natural Regeneration to Restore Degraded Tropical Forestlands. Restoration Ecology, 15(4): 620–626.

[11] Elliott S, Navakitbumrung P, Kuarak C, Zankum S, Anusarnsunthorn V, Blakesley D, 2003. Selecting framework tree species for restoring seasonally dry tropical forests in northern Thailand based on field performance. For Ecol Manage 184:177-191

[12] Goosem, S. and N. I. J. Tucker, 1995. Repairing the Rainforest. Wet Tropics Management Authority, Cairns, Australia. Pp 72. http://www.wettropics.gov.au/media/med_landholders.html

13.6 External links

- The Global Partnership on Forest and Landscape Restoration - More information on global initiatives to restore forest ecosystems

- Issuance Of Eco-Restoration License

Chapter 14

Forests for the 21st Century

Forests for the 21st Century is a short video promoting and explaining the benefits of forest landscape restoration. In the last few centuries people have removed more than half of the world's forest cover.*[1] Deforestation is currently responsible for nearly 20 per cent of global carbon emissions.*[2] This tide of deforestation can be reversed, but we can make a much greater impact if we also put back some of our lost forests. Planting more trees can lock up more carbon, improve the environment and people's lives. Many regions and countries have already restored much of their forest.

14.1 Summary

The film includes examples of successful restoration projects in the UK, at Shinyanga in northern Tanzania and in the Miyun area of China. New information indicates that there are many millions more hectares of lost forests and degraded lands that are suitable for restoration than previously estimated.*[3] The extent of global opportunities for forest restoration is illustrated in the video for the very first time.

The 15 minute film was produced in November 2009 by the Forestry Commission of Great Britain and the IUCN on behalf of the Global Partnership on Forest Landscape Restoration.

14.2 References

[1] IPCC 4th Assessment Report

[2] IPCC - Climate Change 2007

[3] GPFLR - A World of Opportunity

14.3 External links

- Complete film on YouTube

Chapter 15

Futuro Forestal S.A.

Futuro Forestal S.A. is a German-Panamanian reforestation company that operates in Latin America. It was founded 1994 in Panama and headquartered in Panama City. Futuro Forestal is the first impact investment management company of the tropical forestry industry.[1] To date the company has planted over 8,000 hectares of teak and mixed hardwood plantations on deforested pastureland,[2] often under the Forest Stewardship Council (FSC) standard.[3]

15.1 History

Possible Impact of Investments: From Triple bottom line to Impact Investment

Futuro Forestal was founded 1994 by Andreas Eke and Iliana Armién. Since that time, the company developed from a small retail investment to a timber investment management organization to an impact forestry company with up to 2,500 employees.[4][5]

- 1994: Futuro Forestal started its first reforestation, called "Projecto Madera Fina" (engl.: fine timber

project), with 9 hectares in Panama.[6]

- 1998: As first company in Panama Futuro Forestal applies to FSC-standard.[7]

- 1998: Futuro Forestal transferred as first company worldwide a Business-to-business-transaction from reforestation to Carbon credit market.[8]

- 2001: Opening of new offices and a nursery in Las Lajas, Chiriquí, Panama.

- 2003 & 2004: The reforestations of Futuro Forestal were rated as Latin America's best forest investment[9] by rating agency SICIREC (abbr., span.: Sistemas de Circulación Ecológica, engl.: systems of ecological cycles).

- 2005: 2005: Futuro Forestal brings its first shipment of FSC-certified timber to the market. Referring to Jagwood+, the sale of timber from teak and yellow cedar brought a significantly higher price (US $120/m³) than "[...] uncertified thinning wood (normally around 50-70 US$)[10]".

- 2006: Metafore Innovation Award for Futuro Forestal's "[···] WoodStockInvest program, which offers worldwide investors the opportunity to own a forest, invest in a high yield product and contribute to social and ecological development in Central America".[11]

- 2006/2007:Futuro Forestal expanded its operations to Nicaragua and started a reforestation program in cooperation with the United Nations Framework Convention on Climate Change[12] (UNFCCC).

- 2008: Futuro Forestal sold its retail-investment reforestations to its long-term sub-contractor Forest Finance. According to Forest Finance the reforestation-areas of Futuro Forestal brought significantly higher payoff than expected in revenue forecasts.[13]

- 2009: German Investment Corporation (German: Deutsche Investitions- und Entwicklungsgesellschaft, abbr.: DEG) and Futuro Forestal started an environmental education initiative for primary schools in Nicaragua.[*][14]

- 2011: The Company decided to refocus its efforts as an impact investment management company. To make it transparent to its stakeholders, Futuro Forestal reaffirmed its support of the Ten Principles of the United Nations Global Compact.[*][15]

- 2012: The Global Exchange for Social Investment (GEXSI) and Futuro Forestal established a strategic partnership. Through it, Futuro Forestal's experience and methodology will be adopted for an upcoming timber-project in Madagascar.[*][1]

15.2   Services

Futuro Forestal provides sustainable reforestation services including timber investment management, eco system restoration, Corporate Social Responsibility (CSR) project execution and social services like education for rural communities.

2 year old reforestation of Futuro Forestal with teak.

15.2.1   Timber investment

Futuro Forestal developed plantations where native hardwood species like Roble Coral (*Amazonia terminalia*), Cocobolo (*Dalbergia retusa*) and Zapatero (*Hiernonyma alchorneoides*) are planted according to site-specific soil conditions. The philosophy was also applied to exotic teak (*Tectona grandis*). This helps optimize soil conditions and creates multi-faceted habitats for wildlife, particularly

when compared to monoculture plantations. In Futuro Forestal's plantations a significant percentage of the land areas is allocated to environmental protection. Furthermore "[...] the company has been the first to sell carbon credits from reforestation as a business in Panamá" (Montagnini 2005, p. 181).

14-year-old reforestation of Futuro Forestal with native species.

15.2.2   Ecosystem restoration

Ecosystem restoration is the return of a damaged ecological system to a stable, healthy, and sustainable state. Futuro Forestal was awarded an important mitigation project for Minera Panama.[*][16] The 7,000+ hectare project aims to:

- Restore degraded land using angiosperms and pioneer species,

- Establish species natural to the area,

- Optimize ratio and distribution of species used,

- Promote wildlife function through the establishment of trees with food functions, and

- Provide significant biodiversity with the support of experimental nurseries with native species.

15.2.3   CSR project execution

In view of CSR projects is solely the sustainability of reforestation's social and environmental impact.

15.2.4   Education

Together with Nicaraguan Federal Ministry for Economic Cooperation and Development Futuro Forestal started environmental education in forest-dependent communities.

Children are educated in primary schools and adults get theoretical knowledge about agroforestry, like sheep-farming with Pelibüeys and beekeeping. Practical development is generated through microcredits for sheep and beehives.*[17]

15.2.5 Scientific cooperation

In 2001 Futuro Forestal, the Native Species Reforestation Project of the Yale University's School of Forestry and Environmental Studies & the Smithsonian Tropical Research Institute conducted a native species project in Panama. Through this, the partners researched native species silvicultural*[18] and practical application of forest management techniques.*[19]

15.3 Subsidiary

Forest & Community Foundation (nonprofit)

15.4 References

15.4.1 Notes

[1] Global Exchange for Social Investment

[2] Montagnini 2005, p. 181

[3] World Wildlife Fund: http://www.gftn.panda. org/gftn_worldwide/asia/lao_pdr/?19126/ Panamanian-forest-company-commits-to-FSC

[4] *Forest Stewardship Council. La Expereriencia de Futuro Forestal, Managua, Nicaragua.* Slide 09. Retrieved 2014-03-03.

[5] Asociacion Nacional de Reforestadores y Afines de Panama: http://www.anarap.com/?p=254

[6] Official website: http://www.futuroforestal.com/ reforestation/experienc/timeline

[7] FSC-Certificate of Futuro Forestal from 1998 forestfinance.de (PDF-File) Retrieved 2014-02-24

[8] *Natsource Creates Environmental Action Desk to Serve Retail Emissions Trading Market; 'Virtual' Desk Kicks Off with Greenhouse Gas and SO2 Trades.* Business Wire, June 7th, 2000. Retrieved 2014-02-28

[9] "Green Investing, Cleantech Investing, Renewable Energy Investing". Sustainablebusiness.com. Retrieved 2014-03-02.

[10] Jagwood+ http://data2.blog.de/media/377/1161377_ 7d8f9dadd3_d.pdf

[11] "Metafore Announces Recipients of Inaugural Innovation Awards". Marketwired.com. 2006-05-15. Retrieved 2014-03-02.

[12] United Nations Clean Development Mechnism http://cdm.unfccc.int/ProgrammeOfActivities/Validation/DB/ 842J7KCG5SRVNU5AZ2O6DK8SK18ND1/view.html

[13] Press release of Forest Finance Service GmbH: http://www. lifepr.de/pressemitteilung/forest-finance-service-gmbh/ Futuro-Forestal-uebertrifft-Ertragsprognose/boxid/15546

[14] German Federal Ministry for Economic Cooperation and Development (2012), "Resource Efficiency: A Strategy for the Future". develoPPP.report – Magazine for Development Partnerships (32). Access on 010714 under http://www.futuroforestal.com/wp-content/uploads/ develoPPP-report-32-ResourceEfficiency_EN.pdf

[15] Reporting of The UN Global Compact: http://www. unglobalcompact.org/COPs/learner/15240

[16] Directory of Members of Panama Mining Chamber http:// www.camipa.org/directorio.html

[17] Miller, A.M. (2011). "Special Focus on Forests". business .2020, 6(2). Access on 021814 under http://www.cbd.int/ doc/newsletters/B-03800.pdf

[18] "futuro_forestal". Prorena.research.yale.edu. Retrieved 2014-03-02.

[19] "Smithsonian Tropical Research Institute". Stri.si.edu. Retrieved 2014-03-02.

15.4.2 Further reading

Montagnini, Florencia; Jordan, Carl F. (2005). *Tropical Forest Ecology. The Basis for Conservation and Management.* Berlin: Springer. ISBN 3-540-23797-6.

15.5 External links

- Official website

Chapter 16

Gap dynamics

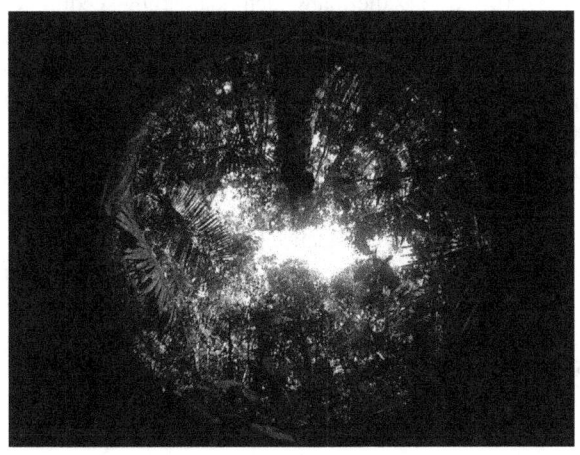

Treefall gaps in the Amazon allow sunlight to reach the forest floor.

Gap dynamics refers to the pattern of plant growth that occurs following the creation of a forest gap, a local area of natural disturbance that results in an opening in the canopy of a forest. Gap dynamics are a typical characteristic of both temperate and tropical forests and have a wide variety of causes and effects on forest life.

Gaps are the result of natural disturbances in forests, ranging from a large branch breaking off and dropping from a tree, to a tree dying then falling over, bringing its roots to the surface of the ground, to landslides bringing down large groups of trees. Because of the range of causes, gaps therefore have a wide range of sizes, including small and large gaps. Regardless of size, gaps allow an increase in light as well as changes in moisture and wind, leading to differences in microclimate conditions compared to those from below the closed canopy, which are generally cooler and more shaded.

For gap dynamics to occur in naturally disturbed areas, either primary or secondary succession must occur. Ecological secondary succession is much more common and pertains to the process of vegetation replacement after a natural disturbance. Secondary succession results in second-growth or secondary forest, which currently covers more of the tropics than old-growth forest.

Since gaps let in more light and create diverse microclimates, they provide the ideal location and conditions for rapid plant reproduction and growth. In fact, most plant species in the tropics are dependent, at least in part, on gaps to complete their life cycles.*[1]

16.1   Disturbances

Main article: Disturbance (ecology)

Gap dynamics are the result of disturbances within an ecosystem. There are both large scale and small scale disturbances, and both are influenced by duration and frequency. These all affect the resulting impact and regeneration patterns of the ecosystem.

The most common type of disturbance within a tropical ecosystem is fire. Since most nutrients in a tropical ecosystem are contained in the biomass of plants, fire is an important component of recycling these nutrients and therefore regenerating an ecosystem.

An example of a small scale disturbance is a tree falling. This can cause soil movement, which redistributes any nutrients or organisms that were attached to the tree. The tree falling also opens up the canopy for light entrance, which can support the growth of other trees and plants.

After a disturbance, there are several ways in which regeneration can occur. One way, termed the advance regeneration pathway, is when the primary understory already contains seedlings and saplings. This method is most common in the Neotropics when faced when small scale disturbances. The next pathway is from tree remains, or any growth from bases or roots, and is common in small disturbance gaps. The third route is referred to as the soil seed bank, and is the result of germination of seeds already found in the soil. The final regeneration pathway is the arrival of new seeds via animal dispersal or wind movement. The most critical components of the regeneration are seed

Broken trees create gaps in the central Amazon.

distribution, germination, and survival.[*][1]

16.2   Forest gaps and forest regeneration

Until recently, forest regeneration practices in North America have largely followed an agricultural model, with research concentrated on techniques for establishing and promoting early growth of planted stock after clearcutting (Cleary et al. 1978, Lavender et al. 1990, Wagner and Colombo 2001),[*][2][*][3][*][4] followed by studies of growth and yield emphasizing single-species growth uninfluenced by overstorey canopy. Coates (2000)[*][5] questioned this approach and proposed a shift to a more ecologically and socially based approach able to accommodate greater diversity in managed stands. Predictive models of forest regeneration and growth that take account of variable levels of canopy retention will be needed as the complexity of managed forest stands increases (Coates 2000).[*][5]

Tree regeneration occurring inside canopy gaps after disturbance has been studied widely (Bazzaz and Pickett 1980,

Platt and Strong 1989).[*][6][*][7] Studies of gap dynamics have contributed much to an understanding of the role of small-scale disturbance in forest ecosystems, but they have been little used by foresters to predict tree responses following partial cutting (Coates and Burton 1997).[*][8]

In high-latitude northern forests, position inside a gap can have a pronounced effect on resource levels (e.g., light availability) and microclimate conditions (e.g., soil temperature), especially along the north–south axis. Such variation must inevitably affect the amount and growth of regeneration; but relying solely on natural regeneration to separate the effects of gap size and position is problematic (Coates 2000).[*][5] Among the many factors affecting seedling establishment following canopy disturbance are parent tree proximity and abundance, seedbed substrate, presence of seed consumers and dispersers, and climatic and microclimatic variability. Planted trees can be used to avoid many of the stochastic events surrounding natural seedling establishment.

Gradients of canopy influence can be created by partial cutting, and tree growth responses within gaps of various sizes and configurations, as well as within the adjacent forest matrix can form a basis for tree species selection. Hybrid spruce (the complex of white spruce, Sitka spruce, and occasionally Engelmann spruce) was one of several coniferous species used in a study in the Moist Cold subzone of the Interior Cedar–Hemlock zone in northwestern British Columbia. A total of 109 gaps were selected from a population of openings created by logging within each light and heavy partial cutting treatment in stands averaging 30 m in canopy height; 76 gaps were less than 1000 m^2, 33 were between 1000 m^2 and 5000 m^2. Canopy gap size was calculated as the area of an ellipse, the major axis of which was the longest line that could be run from canopy edge to canopy edge inside the gap, and the minor axis was the longest line that could be run from canopy edge perpendicular to the long line. Seedlings were planted in gaps and in the undisturbed and clearcut treatment units. There were strong and consistent trends in growth response among the seedlings as gap size increased. In all species, growth increased rapidly from small single-tree gaps to about 1000 m^2, but thereafter, there was little change up to 5000 m^2. Tree size and current growth rates for all species were highest in full open conditions. In large and medium gaps (300–1000 m^2), the largest trees of all species occurred in the middle gap position, with little difference between the sunny north and shady south positions, lodgepole pine excepted. The light advantage expected off the north end of higher-latitude gaps was not a benefit for tree growth, suggesting that below-ground effects of canopy edge trees have an important influence of seedling growth in these forests (Coates 2000).[*][5]

In a study near Chapleau, Ontario (Groot 1992, Groot et al.

1997),[*][9][*][10] openings were created in 40-year-old aspen and monitored to determine their influence on outplanted white spruce seedling development. Circular openings 9 m and 18 m in diameter, 9 m and 18 m wide east–west strips, and a 100 m × 150 m clearcut were planted and spot-seeded. The variation in solar radiation, air temperature, and soil temperature among the strips and plots was almost as great as the variation between the clearcut and intact forest. Solar radiation during the first growing season varied from 18% of the above-canopy values within the uncut stand to 68% values at the center of the 18 m strip. Near the edges of the strips, solar radiation was about 40% of the above-canopy along the south and 70% to 80% along the north. Stomatal conductance in white spruce seedlings declined generally from more sheltered to more exposed environments, correlating best with increased vapor pressure deficit (VPD). Without vegetation control, position in openings had little effect on the growth of planted white spruce; regrowth of lesser vegetation isolated seedlings from the microclimatic effects of overstorey treatment. Seedling diameters were independent of environment, while height growth was only slightly greater in environments having more light. With vegetation control, white spruce diameter and height were greatest in the center of the strips, even though there was less light there than along the north edge of the strips. Moisture stress may have accounted for that result.

Cecropia trees are a common pioneer species found in gaps.

16.3   Primary succession

Succession is the slow rebuilding of forest gaps from natural or human disturbances. When major geological changes such as volcano eruptions or landslides occur, the current vegetation and soil may erode away leaving only rock. Primary succession occurs when pioneer species such as lichens colonize rock. As the lichens and mosses decompose, a soil substrate forms called peat. The peat, over time, will create a terrestrial ecosystem. From there on herbaceous, non-woody plants will develop and trees will follow. Major holes or gaps in the forest ecosystem will take hundreds of years to regenerate from a rock base.[*][11]

16.4   Secondary succession

Secondary succession occurs where a disturbance has taken place but soil remains and is able to support plant growth. It does not take nearly as long for plant regeneration to occur because of the soil substrate already present. Secondary succession is much more common than primary succession in the tropics.

Ecological secondary succession occurs in four distinct phases: First, rapid colonization of cleared land by species such as herbs, shrubs, and climbers as well as seedlings from pioneer tree species occurs and this can last up to three years. After that, short lived but fast growing shade intolerant species form a canopy over 10 to 30 years. Non-pioneer heliophilic (sun-loving) tree species then add to the biomass and species richness as well as shade tolerant species and this can last 75 to 150 years. Finally, shade-tolerant species regain full canopy stature indefinitely until another major disturbance occurs.[*][12]

Secondary succession in the tropics begins with pioneer species, which are rapidly growing and include vines and shrubs. Once these species are established, large heliophilic species will develop such as heliconias. Cecropias are also a major pioneering tree in the tropics and they are adapted to grow well where forest gaps are giving way to sunlight. Shade-tolerant species that have remained low in the forest develop and become much taller. These successional phases do not have definite order or structure and because of the very high biodiversity in the tropics, there is a lot of competition for resources such as soil nutrients and sunlight.

16.5 Examples of tree dynamics

Due to the fact that horizontal and vertical heterogeneity of a forest is significantly increased by gaps, gaps become an obvious consideration in explaining high biodiversity. It has been proven that gaps create suitable conditions for rapid growth and reproduction. For example, non-shade tolerant plant species and many shade-tolerant plant species respond to gaps with an increase in growth, and at least a few species are dependent on gaps to succeed in their respected environments (Brokaw 1985; Hubbell and Foster 1986b; Murray 1988; Clark and Clark 1992). Gaps create diverse microclimates, affecting light, moisture, and wind conditions (Brokaw 1985). For example, exposure to edge effects increases a microclimate's light and wind intensity and decreases its moisture. A study conducted on Barro Colorado Island in Panama showed that gaps had greater seedling establishment and higher sapling densities than control areas.

Species richness was higher in gaps than in control areas, and there was more diversity in species composition among gaps. However, this study also found that there was a low recruitment rate per gap, which explains why gaps differed in species composition. With 2% to 3% for pioneer species and 3% to 6% for shade-tolerant and intermediate species. Suggesting that most species could not take advantage of gaps because they couldn't get to them through seed dispersal. With that said, the Janzen-Connell effect plays a major role in the tree species' relationship with gaps. The Janzen-Connell density dependent mortality model states that most trees die as seed or seedlings. In addition, host-specific predators or pathogens are predicted to be greatest where density is greatest, which is underneath parent tree. This corroborates with the major causes of gaps, which are the falling of trees due to mortality caused by termites or epiphyte growth. The Janzen-Connell model also states that balance between dispersal distance and mortality should cause highest recruitment to be at a certain distance away from the parent. Therefore if these gaps are being created by the parents, the seedlings recruit away from the gap, resulting in increasing survival rates as the distance from the parent increases. This explains the low recruitment rate per gap found in the experiment conducted in Barro Colorado Island.*[13]

In corroboration, a study conducted in La Selva in Costa Rica calculated the crown illumination index for nine tree species ranging from gap specialists to emergent canopy species. Crown illumination values ranged from 1, which indicated low light, and 6, which indicated that the tree crown was completely exposed . After using a mathematical model to calculate the changes in tree diameter and changes in crown illumination with age. This model helped estimate life expectancy, time of passage to various sizes, and age patterns of mortality. The results showed what most

gap dynamics studies show, pioneer species thrived in high light environments and non-pioneer species showed high mortality when young but the rate of mortality decreased as they aged. However, once trees were very large survivorship then decreased.*[14]

16.6 References

[1] Kricher, John (2011). *Tropical ecology*. Princeton, New Jersey: Princeton University Press. pp. 188–226.

[2] Cleary, B.D.; Greaves, R.D.; Hermann, R.K. (Compilers and Eds.). 1978. Regenerating Oregon's Forests. Oregon State Univ. Exten. Serv., Corvallis OR. 287 p.

[3] Lavender, D.P.; Parish, R.; Johnson, C.M.; Montgomery, G.; Vyse, A.; Willis, R.A.; Winston, D. (Eds.). 1990. Regenerating British Columbia's Forests. Univ. B.C. Press, Vancouver BC. 372 p.

[4] Wagner, R.G.; Columbo, S.J. (eds.). 2001. Regenerating the Canadian forest: Principles and practice for Ontario. Fitzhenry & Whiteside, Markham, Ont.

[5] Coates, K.D. 2000. Conifer seedling response to northern temperate forest gaps. For. Ecol. Manage. 127 (1–3):249–269.

[6] Bazzaz, F.A.; Pickett, S.T.A. 1980. Physiological ecology of tropical succession: A comparative review. Ann. Rev. Ecol. Syst. 11:287–310.

[7] Platt, W.J.; Strong, D.R. 1989. Special feature: gaps in forest ecology. Ecology 70:535–576.

[8] Coates, K.D.; Burton, P.J. 1997. A gap-based approach for development of silvicultural systems to address ecosystem management objectives. For. Ecol. Manage. 99:337–354.

[9] Groot, A. 1992. Small forest openings to promote the establishment and growth of white spruce in boreal mixedwood stands. Draft NODA proposal, with comments by R.F. Sutton.

[10] Groot, A.; Carlson, D.W.; Fleming, R.L.; Wood, J.E. 1997. Small openings in trembling aspen forest: microclimate and regeneration of white spruce and trembling aspen. Nat. Resour. Can., Can. For. Serv., Sault Ste. Marie ON, NODA/NFP Tech Rep. TR-47. 25 p.

[11] Brokaw, N.V.L. (1985). *The Ecology of Natural Disturbance and Patch Dynamics*. San Diego, California: Academic Press. pp. 53–69.

[12] Miguel Martínez-Ramos, Elena Alvarez-Buylla and José Sarukhán (June 1989). "Tree Demography and Gap Dynamics in a Tropical Rain Forest". *Ecology* **70** (3): 555–558. doi:10.2307/1940203.

[13] Hubbell, S. P. and R. B. Foster (1986). *Plant Ecology*. Oxford, UK: Blackwell. pp. 77–95.

[14] Clark, JS (1992). *Ecosystem Rehabilitation: Preamble to Sustained Development.* SPB Academic Publishing. pp. 165–186.

Chapter 17

Green Belt Movement

The **Green Belt Movement (GBM)** is an indigenous grassroots non-governmental organisation based in Nairobi, Kenya that takes a holistic approach to development by focusing on environmental conservation, community development and capacity building. Professor Wangari Maathai established the organisation in 1977, under the auspices of the National Council of Women of Kenya.

The Green Belt Movement organises women in rural Kenya to plant trees, combat deforestation, restore their main source of fuel for cooking, generate income, and stop soil erosion. Maathai has incorporated advocacy and empowerment for women, eco-tourism, and just economic development into the Green Belt Movement.

Since Maathai started the movement in 1977, over 51 million trees have been planted. Over 30,000 women trained in forestry, food processing, bee-keeping, and other trades that help them earn income while preserving their lands and resources. Communities in Kenya (both men and women) have been motivated and organised to both prevent further environmental destruction and restore that which has been damaged.

In 2004, Wangari Maathai received the Nobel Peace Prize – becoming the first African woman to win the Nobel Peace Prize – for her work with the Green Belt Movement. Her book, *The Green Belt Movement* is published by Lantern Books. Maathai was a leader in ecofeminist movement.

17.1 The movement

17.1.1 Structure

There are two divisions of the Green Belt Movement: Green Belt Movement Kenya (GBM Kenya) and the Green Belt Movement International (GBMI).

Key focus areas

The Green Belt Movement works in six principal areas, known as "core programs":

- Advocacy & Networking

- Civic & Environmental Education

- Environmental Conservation/Tree Planting

- Green Belt Safaris (GBS)

- Pan African Training Workshops; and

- Women for Change (capacity building)

Each of these programs is aimed at improving the lives of local inhabitants by mobilising their own abilities to improve their livelihoods and protect their local environment, economy and culture.

17.1.2 History

In 1972, the environmental movement revolutionised advocacy and policies surrounding environmental issues such as those in The United Nations Environment Programme, also known as (UNEP). UNEP was established in Nairobi as a result of the United Nation Conference on Human Environment held in Stockholm in the same year. This development helped arouse interest in the environment in Africa regardless of the fact that many governments in the region held hostile sentiments towards the policies adopted in Stockholm to limit environmental degradation. Soon after, Maathai served as chairwoman of the UNEP's Environment Liaison Center board, which today is called the Environment Liaison Center International. In 1974, Maathai's focus became forestation and reforestation issues. She introduced a tree-planting program and opened the first tree nursery, from which she formed Envirocare Ltd. Although this program experienced many setbacks because of a lack of funding and support,*[1] it facilitated Maathai's involvement with the National Council of Women of Kenya as a member of the Executive Committee in 1977. Her determination to inexpensively provide the rural women of the NCWK with sufficient wood for fuel, building, and soil conservation, inspired the Save the Land Harambee tree-planting initiative.*[2] This soon began a widespread tree-planting strategy in which over a thousand seedlings were planted in long rows to form green belts of trees, and thus marking the very beginning of the Green Belt Movement.*[3]

> "These "belts" had the advantages of providing shade and windbreaks, facilitating soil conservation, improving the aesthetic beauty of the landscape and providing habitats for birds and small animals. During these local tree-planting ceremonies, community members usually turned out in large numbers. To conceptualise this fast-paced activity of creating belts of trees to adorn the naked land, the name Green Belt Movement was used." *[4]

From 1977–1988, the movement steers clear of traditional political arenas seeking to transform the social ground through reforestation and education. During the second phase, 1989–1994, the Green Belt Movement maintains these non-confrontational goals, while Wangari Maathai openly challenges the political arena. Throughout the Green Belt Movement, the organizers have been conscientious in framing their beliefs in a non-violent way. As a result, consensus, and not conflict or disruption among environmental issues has been the catalyst for major change in the social and political arena.*[5]

17.2   Projects

- 1980s: Establishment of over 600 tree nurseries achieved (2,500 – 3,000 women assisting)

- 1980s: Establishment of approximately 2,000 public green belts carrying 1,000 tree seedlings on each green belt

- mid-1980s: Pan-African Green Belt Network developed (since adopted in Tanzania, Uganda, Lesotho, Malawi, Zimbabwe etc.)

- 1988: Struggle against construction of Africa's tallest skyscraper in Uhuru Park Nairobi (see "Activism Against the Odds" below)

- 2008: Support of the Billion Tree Campaign

17.3   Activism

In 1989 the Movement took on the powerful business associates of President Daniel arap Moi. A sustained, and often lonely protest, against the construction of a 60-story business complex in the heart of Uhuru Park in Nairobi was launched and won.

In 1991 a similar protest was launched that saved Jeevanjee Gardens from the fate of being turned into a multi-story parking lot.

In 1998, the Movement led a crusade against the illegal allocation of parts of the 2,000 acre (8 km^2) Karura Forest, a vital water catchment area in the outskirts of Nairobi. The struggle was finally won in 2003 when leaders of the newly elected NARC government affirmed their commitment to the forest by planting trees in the area.

This activism has come at a high cost to both Maathai in person and to the Movement. The Kenyan government closed Greenbelt offices, has twice jailed Maathai and she was subject in 1992 to a severe beating by police while leading a peaceful protest against the imprisonment of several environmental and political activists. Whilst these have served as impediments to the Greenbelt Movement, they have not stifled it and it continues as a world-renowned and respected Movement.

In 2007, the Green Belt Movement endorsed the Forests Now Declaration, calling for new market based mechanisms to protect tropical forests.

17.4   Future prospects

In the early 21st century, the Movement is now vibrant and has succeeded in achieving many of the goals it set out to meet. Environmental protection has been achieved through tree planting, including soil conservation, sustainable management of the local environment and economy and the protection and boosting of local livelihoods. In addition to helping local women to generate their own incomes through such ventures as seed sales, the Movement has succeeded in educating thousands of low-income women about forestry and has created about 3,000 part-time jobs. The movement also aims to spreading its roots to all the countries in the world

17.5   See also

- Lantern Books

- Marion Institute

- SeedTree*[6]

17.6 References

[1] The Green Belt Movement, Wangari Maathai, 2006

[2] The Green Belt Movement, Wangari Maathai, 2006

[3] The Green Belt Movement, Wangari Maathai, 2006

[4] The Green Belt Movement, Wangari Maathai, 2006

[5] Michaelson, M. Wangari Maathai and Kenya's Green Belt Movement: Exploring the Evolution and Potentialities of Consensus Movement Mobilization, 1994

[6] SeedTree

17.7 External links

- The Green Belt Movement

o all the countries in the world

17.8 See also

- Lantern Books
- Marion Institute

17.9 References

17.10 External links

- The Green Belt Movement

Chapter 18

Groasis Waterboxx

The Groasis Waterboxx

The **Groasis Waterboxx** is a device designed to help grow trees in dry areas. It was invented and developed by Dutch former flower exporter Pieter Hoff,[1] and won the Popular Science Green Tech Best of What's New Innovation of the year award for 2010.[2][3]

18.1 Background

Large land areas in the world are too dry for trees to survive. Although water may be present in the ground, it is often too deep for small trees to develop a root structure to reach.[3] The Groasis Technology employs biomimicry[4] to solve the problem of growing plants in deserts, eroded areas, badlands and on rocks. The purpose of this technology is to replant such areas, restore the vegetation cover and make them productive with fruit trees and vegetables.

18.2 Design

The Groasis is a polypropylene bucket with a lid.[5] It has a vertical tunnel in the middle for two plants. A wick allows water from inside the box to trickle into the ground via capillary action. The device mimics the insulating effect bird feces provide to germinating seeds.[6][7] The box's lid is covered by tiny papillae, which create a superhydrophobic surface due to the lotus effect. The lid

serves to funnel even the smallest amount of water down siphons into the box's central reservoir.[8][9]

The product functions as a plant incubator, sheltering both a newly planted sapling and the ground around it from the heat of the sun, while providing water for the plant. The lid collects water from rain and nighttime condensation, which is then stored in the bucket. The water-filled reservoir releases small amounts (around 50 ml per day) of water into the ground by a wick to water the tree and to encourage the tree to develop a root structure.[3] The box acts as a shield for the water in the upper ground, and this water then spreads down and out instead of being drawn to the surface and evaporated.[3] Both temperature and humidity beneath and inside the box are more stable night and day than without.[7]

As of 2010, the development has taken 7 years at a cost $7.1 million.[10]

18.3 Installation

Use of the box initially involves digging a hole in the ground by a human or a machine. One to three plants are planted in the hole, and a cardboard panel is placed around the plants. In dry areas, the soil around the plants is inoculated with mycorrhizae to release nutrients in the soil that would otherwise be chemically inaccessible to the growing plants. A wick is inserted in the bottom of the Groasis which is then lowered over the plants and filled with water. Two lids are put on, funnels inserted and a cap plugs the top lid.[11]

18.4 Testing

The box has been tested for 3 years at Mohamed Premier University in Morocco where nearly 90% of plants survived with the box compared to 10% without.[12][13] Apart from projects in warm arid areas, the box is being tested in wineries and cold mountain regions.[14][15][16][17]

The device is also being used to grow water-loving trees in temperate regions, including growing giant sequoia (*Sequoidendron giganteum*) in the Great Lakes region.

18.5 See also

- Agroforestry
- Applied ecology
- Desertification
- Ecological engineering methods
- Environmental technology
- Green Sahara
- Air well (condenser)

18.6 References

[1] Witkin, Jim. Developing a 'Water Battery' for trees *New York Times*, 9 April 2010. Accessed: 5 December 2010.

[2] Jannot, Mark. Best of What's New 2010: Our 100 Innovations of the Year *Popular Science*, 16 November 2010. Accessed: 5 November 2010.

[3] AquaPro Holland Groasis Waterboxx *Popular Science*. Accessed: 5 December 2010.

[4] Susan Kraemer, "Inventor Uses Biomimicry To Create Dew", *Cleantechnica.com*

[5] Parsons, Sarah. Groasis Waterboxx can grow trees in any climate – even the desert *Inhabitat*, 4 December 2010. Accessed: 5 December 2010.

[6] Buczynski, Beth. New tree-growing device inspired by bird poop *Care2*, 30 November 2010. Accessed: 5 December 2010.

[7] Coxworth, Ben. Groasis Waterboxx lets trees grow up in unfriendly places *GizMag*, 18 November 2010. Accessed: 5 December 2010.

[8] http://dewharvest.blogspot.com/2014/01/the-lotus-leaf-inspired-waterboxx-lid.html

[9] http://www.groasis.com/en/technology/the-different-forms-of-condensation

[10] Binns, Corey. Invention Awards: A box that keeps plants hydrated in the desert *Popular Science*, 25 May 2010. Accessed: 6 December 2010.

[11] A'Hearn, Peter. Groasis Waterbox tree planting demo (Video) *TeacherTube*, 20 September 2010. Accessed: 5 December 2010.

[12] Fernandes, Sunil. Oil & Gas page 34-36 *Oil & Gas Review*, May 2010. Accessed: 5 December 2010.

[13] Growing trees in the desert, with the aid of a 'Waterboxx' *Voice of America*, 12 August 2010. Accessed: 5 December 2010.

[14] Thinking inside the Groasis Waterboxx solves deforestation, water depletion, food shortage *PR Newswire*, 22 June 2010. Accessed: 5 December 2010.

[15] Kasica, Stephen. Eagle River gets restoration tips from the Sahara *Vail Daily* 23 May 2012. Retrieved: 6 June 2012.

[16] New Tree Seedlings Planted Along North Austin Bus Routes 30 March 2012. Retrieved: 6 June 2012.

[17] Waterboxx experiment *Sustainable Neighborhoods of North Central Austin* 23 May 2012. Retrieved: 6 June 2012.

18.7 External links

- Official website
- University of Valladolid starts in Spain the 2 billion hectares reforestation project with Groasis on YouTube
- How does the Groasis waterboxx work against desertification? on YouTube
- The Arid Arborist: A blog about results with Groasis Waterboxx

Chapter 19

Hauberg

This article is about a type of forest management. For the rare name of a type of German farmhouse, also spelt *Hauberg*, see Haubarg.

Hauberg (also **Hackberg**) is a type of communal forest

The Hauberg near Netphen

Woodpile near Salchendorf

management that is typical of the Siegerland and adjacent parts of the Lahn-Dill Uplands and the Westerwald in central Germany. Its aim is to manage the forest in order to produce tanbark and charcoal for the regionally important

iron ore industry as well as firewood. In addition to forestry uses, the area also has agricultural uses, such as the growing of rye and buckwheat, typical of shifting cultivation, in the year after the timber harvest, as well as subsequent communal grazing (commons).

19.1 Overview

The Hauberg is an oak-birch coppiced woodland, in which other trees are occasionally scattered. With a cycle of from 16 to 20 years the Hauberg undergoes clearcutting or coppicing, leaving the stumps in the ground to begin growing again. Only in the year after clearfelling is the land used for grain. In years when there is a lot mast, pigs are kept in the Hauberg.

With the decline in demand for tanbark and charcoal this form of management has losts its significance. In the second half of the 20th century large areas were turned over to high forest management. The remaining low forest stands are almost exclusively devoted to the production of firewood and industrial wood.

19.2 See also

- Haubarg or Hauberg, a typical farmhouse type on the Eiderstedt peninsula

19.3 Literature

19.3.1 General sources

- Becker, Alfred: ˮDer Siegerländer Haubergˮ . Vergangenheit, Gegenwart und Zukunft einer Waldwirtschaftsform. verlag die wielandschmiede, Kreuztal, 1991

- Becker, Alfred: Das Haubergs-Lexikon. verlag die wielandschmiede, Kreuztal, 2002

- Hans Hausrath: *Geschichte des deutschen Waldbaus. Von seinen Anfängen bis 1850.* Schriftenreihe des Instituts für Forstpolitik und Raumordnung der Universität Freiburg. Hochschulverlag, Freiburg im Breisgau, 1982, ISBN 3-8107-6803-0

- Richard B. Hilf: *Der Wald. Wald und Weidwerk in Geschichte und Gegenwart - Erster Teil* [Reprint]. Aula, Wiebelsheim, 2003, ISBN 3-494-01331-4

- Josef Lorsbach: "Hauberge und Hauberggenossenschaften des Siegerlandes". Band X Quellen und Studien des Instituts für Genossenschaftswesen an der Universität Münster. Verlag C.F. Müller, Karlsruhe, 1956. (Dissertation 1953)

19.3.2 Specific sources

- Konrad Fuchs: *Geschichte der Verbandsgemeinde Gebhardshain 1815-1970.*, Mainz, 1982, ISBN 3-87439-082-9

- Rolf Lerner: *Haubergsgenossenschaften im Kreis Altenkirchen.*, Verlag Mühlsteyn, Elben-Weiselstein 1993

- Manfred Kohl: *Die Dynamik der Kulturlandschaft im oberen Lahn-Dillkreis - Wandlungen von Haubergswirtschaft und Ackerbau zu neuen Formen der Landnutzung in der modernen Regionalentwicklung.* Gießener Geographische Schriften, Heft 45. 176 pp.(and map material), Gießen, 1978

- Alfred Becker (Red.): *Bilder aus dem Hauberg. Naturschutz außerhalb von Schutzgebieten.* Schriftenreihe der Landesforstverwaltung Nordrhein-Westfalen, Heft 1. 3., unveränderte Auflage. Forstliche Dokumentationsstelle der Landesforstverwaltung NRW, Arnsberg 2003, 48 pp., ISBN 3-9809057-5-6

- Frank Schüssler: Die Haubergswirtschaft: Potenziale und Risiken eines traditionellen forstlichen Betriebssystems. In: Geographische Rundschau 01/2008.

19.3.3 Control aspects

- Suchanek in Herrmann/Heuer/Raupach, Kommentar zur EStG und KStG, § 3 KStG, Anmerkung 35ff.

19.4 Weblinks

- viele Infos und Fotos, auch zum Forum Niederwald

- Hauberge

- Hauberg online

-

- siegerlaender-hauberg.info

- Faltblatt "Haubergswirtschaft im Siegerland" (PDF-Datei; 501 kB)

Chapter 20

Hoedads Reforestation Cooperative

The **Hoedads Reforestation Cooperative** (formally, *Hoedads Cooperative Inc.*) was a worker-owned tree planting and forestry labor cooperative based in Eugene, Oregon, United States. It was active throughout the American West from 1971 to 1994. For several years they were country's largest worker-owned cooperative. They were noted for their success in applying the cooperative model successfully to treeplanting. They were also known for their experimentation with and early embrace of concepts such as environmentalism, feminism and alternative economics.[1]

The Hoedads took their name from their use of the "hoedad" (or "hoedag"),[2] a hand implement similar to a hoe used to plant bare-root trees on steep slopes (Hartzell 1987: 29, 45-46).

20.1 Origins

The Hoedads were started by Jerry Rust (later a Lane County Commissioner) and John Sundquist.[3] Rust had returned from a Peace Corps stint in India in the late 1960s and found work planting trees.[3] Both Rust and Sundquist had a love of tree planting but realized that the economics of the industry favored those who organized work crews to bid on jobs with the government or forest owners, rather than merely laboring. They organized their first work unit in 1971, and successfully bid on reforestation projects, beginning with a subcontract in the Tiller District of the Umpqua National Forest (Hartzell 1987). The eventual success of the early Hoedads crew was such that by the late summer of 1973, the group was ready to expand. A meeting was called near Eugene, Oregon which attracted nearly 200 interested workers. Gary Rurkun, writing in the *Whole Earth Catalog*, explained how the coop expanded:

> "A woodsy type character with a full beard, and huge build, and powerful voice called us all around him to chat...He and his friends standing next to him had planted trees as a cooperative

crew for a few years already and had really enjoyed working together, unlike commercial tree planting operations with a crew boss and hourly wage workers. They went on to explain the advantages of a cooperative tree planting for making money as well as for our souls.

"But these boys didn't just want to enjoy their crew, being true new age entrepreneurs, they did not want to give the assembled crowd any work. They offered to get contracts for us, all of us, if we could organize ourselves into cohesive cooperative crews like their crew...They then went on to explain how to organize a crew, the necessity of some money, a crew crummy (the rig in which you ride to work), a treasurer, a crew ideology or by-laws, etc. A lot of information was passed on how they intended for us to work together." [4]

20.2 Economics

The early success of the Hoedads mirrored a unique situation in both American society and in the forestry industry. Generations of clear cutting in the American West had left a huge unmet need for replanting. The U.S. Forest Service, and large landowners, solicited bids to replant trees on a contract basis. Thus work was readily available. The main barrier to entry was to organize a crew, and in many cases, to have land or assets available to pledge as collateral when bidding on a contract. Tree planting itself is hard work, however is not hard to learn and thus motivated crews could quickly be trained and put to work.

The hippie and back to the land movements of the era influenced the Hoedads heavily. There existed large numbers of able bodied young people interested in working the land in a cooperative fashion. Tree planting also lends itself to work in an egalitarian manner; the work is not complex and all planters on a crew can perform the same task.

Many Hoedads had educational backgrounds far beyond

what is typical in the logging industry. For example, co-founder Jerry Rust mentioned his surprise upon realizing everyone in his crew had a college degree,[*][5] and it was not uncommon for crews to have members with post-graduate degrees. Thus many crew members had aspirations beyond merely laboring and actively participated in building the cooperative structure of the Hoedads. However, this also led to a high turnover, as some members grew tired of the repetitive and backbreaking aspects of tree planting work and moved on to other endeavors after a year or two.

20.3 Cooperative structure

The Hoedads' embrace of direct worker democracy led to some long debates and experiments in pay structure. One of the earliest debates was in how members should be paid — by the tree, as was common in the forest industry, or by the hour. Each work crew functioned independently and could set its own policy. Some crews paid by the tree, some by the hour, and some by some combination thereof. Other crews chose "equal pay for equal effort," paying each planter an equal share of the revenue earned by the crew on any particular day.

20.4 Environmentalism and feminism

The Hoedads are credited as being the first group to challenge the notion of forestry work as an all-male domain and most Hoedads work crews included women. Some work crews even were all female.

20.5 End of the cooperative

The Hoedads Reforestation Cooperative ended in 1994 or 1995. In some ways they were a victim of their own success: By the 1990s, growing environmental consciousness led to reduced clearcutting. Fewer contracts for forest replanting were available. At the same time, a growing labor force of migrant Hispanic workers was available and many undocumented workers were willing to work for low wages on commercial tree planting crews.

20.6 References

[1] Roscoe Caron (Late Summer 2001). "Hoedads Celebrate Reforestation History". *West by Northwest Online Magazine.*

Retrieved September 13, 2009. Check date values in: |date= (help)

[2] Steve Nix (28 May 2008). "Hoedads: The Tool, The Cooperative". About.com.

[3] Lois Wadsworth (August 1, 2001). "Tree Planters: The Mighty Hoedads, Back for a 30 year Reunion, Recall Their Grand Experiment". *Eugene Weekly.* Retrieved September 13, 2009.

[4] Stuart Brand, ed. (1980). *The Next Whole Earth Catalog.* Sausalito, CA: Point. ISBN 0-394-73951-5.

[5] Jeff Wright (July 30, 2001). "Back to the Woods, 30 years later". *The Register-Guard* (Eugene). Retrieved October 3, 2009.

20.7 Further reading

- Hartzell, Hal Jr. (1987). *Birth of a Cooperative: Hoedads, Inc. A Worker Owned Forest Labor Co-op.* Eugene, OR: Hulogos'i Communications. ISBN 0-938493-09-4.

- Heilman, Robert Leo (Autumn 2011). "With a Human Face: When Hoedads Walked The Earth". *Oregon Quarterly.*

- Horowitz, Howard. (1986). *Close to the Ground: One Treeplanter's Geography.* Eugene, OR: Hulogos'i Communications. ISBN 0-938493-07-8 (poetry)

20.8 External links

- Hoedads Online, website

- *Green Side Up,* as told by Robert Hirning, YouTube.com

- "Guide to the Hoedads Cooperative Inc. Records," Northwest Digital Archives

- *Oregon Hoedads,* Oregon State University Extension, YouTube.com

Chapter 21

International Small Group and Tree Planting Program

The **International Small Group and Tree Planting Program**, or TIST, is a comprehensive sustainable development program for developing-world locations.[*][1] TIST was started in 2000 and exists in Africa (Kenya, Tanzania, Uganda) and India. The TIST program has three separate but related aims: development, commercial opportunity, and replication.

The development goal of the TIST program is to empower and equip subsistence farmers to restore their natural environment, increase soil fertility, create jobs, strengthen the local community, and move from famine to surplus.

21.1 Program summary

TIST trains and encourages small groups to develop and share "best practices." TIST introduces improved farming and land use techniques to isolated subsistence farmers who are now planting millions of new trees. Using a combination of small group development and training programs and providing small stipends to groups, TIST helps local farmers meet their economic needs, even during severe dry seasons.

Small groups agree to meet the program requirements and assure tree survival and use of improved, sustainable land use techniques for years to come. The improved farming practices and tree planting will improve local welfare by stabilizing the local food supply and by providing families with additional income from TIST tree benefits and payments.

TIST small groups are also educated about HIV/AIDS and equipped to formulate a response to this pandemic at the group and village level. Adopting conservation farming techniques increases food and decreases annual physical effort after the first seedbeds are created. Family members can continue to plant in these seedbeds year after year and have food.

Most of the 60,000 TIST farmer participants were currently using the traditional 3-stone cook stoves or handmade mud stoves. Through work with Envirofit International, TIST has been able to bring healthy cook stoves and the training necessary for an effective replacement for their traditional stoves to many of TIST members.

Project areas

- Project areas

TIST currently operates in four countries:

- India[*][2]
- Kenya[*][3]
- Tanzania[*][4]
- Uganda[*][5]

21.2 TIST origins

The initial pilot implementation site of TIST was in Mpwapwa, Tanzania. Mpwapwa is located southeast of Tanzania's capital, Dodoma.

21.3 References

[1] Scholz, Sebastian M. (2009). "Case study: The International Small Group and Tree Planting Program". *Rural development through carbon finance*. Peter Lang. pp. 27–70. ISBN 978-3-631-59250-2.

[2] "Country profile: India". TIST.

[3] "Country profile: Kenya". TIST.

[4] "Country profile: Tanzania". TIST.

[5] "Country profile: Uganda". TIST.

21.4 External links

- Official website

- USAID Partner TIST: First In The World! http://issuu.com/mitimagazine/docs/miti-12, page 21

- "Farmers form TIST Small Groups. The Small Groups plant trees. Trees create carbon credits. Carbon credits are sold. Farmers make money. The idea is straightforward -- the results are remarkable." http://issuu.com/mitimagazine/docs/miti19, page 10-11

- "Farmers in the Meru and Nyeri areas of the Mt. Kenya Region are ahead of many African farmers in innovations to take advantage of global climate change efforts." Enock W. Kanyanya, USAID Kenya

- http://issuu.com/mitimagazine/docs/miti-7, page 12

- "Rarely do you find a business that attacks two big problems–global poverty and climate change–at the same time." Marc Gunther 2010, http://theenergycollective.com/marcgunther/33010/fighting-poverty-and-global-warming-africa

- TIST has made large, long-term corporate investments in engaging 50,000 Kenyan farmers in a carbon enterprise. This private enterprise model, alongside the donor/philanthropic one, is clearly a valid and useful approach to pioneering forest carbon activities that can bring livelihood benefits to local communities. (page 64) http://theredddesk.org/sites/default/files/resources/pdf/east_africa_study.pdf

- A striking feature of TIST is its scale and rapid growth with more than 50,000 farmer participants and six million trees planted (in Kenya) at the time of the study field work. (page 22) http://www.tist.org/i2/kenyagrowth.php

- Aggregate benefits are the incentive package that motivates carbon sequestration activities in both the immediate and long-term. At current carbon prices, carbon revenues seem insufficient to provide adequate inventive for tree-planting. Co-benefits alone seem to provide sufficient rewards to compensate and exceed tree-planting costs for many, but carbon revenue is important as an organizing principle and a behavioral incentive to each tree-grower. (page 60) http://theredddesk.org/sites/default/files/resources/pdf/east_africa_study.pdf

- New Perimeter provided a range of pro bono legal services to small groups of subsistence farmers in Kenya, Tanzania and Uganda in coordination with the International Small Group and Tree Planting Program (TIST). http://www.newperimeter.org/export/sites/new-perimeter/downloads/main/New-Perimeter-Africa-Overview-Brochure_082013_FINAL.pdf

Chapter 22

Kidney tray (tool)

Kidney tray

A **Kidney tray** is a tool for transporting seedlings.

A kidney tray can be used with shoulder straps and waist straps. Kidney trays are often plastic or a from of plasticized textile can form multiple structures with the right frame.

A twin kidney tray has two trays which are attached to a workers' sides via a shoulder harness and waist strap.

22.1   See also

Pottiputki (tool)

Chapter 23

Theodore Lukens

Theodore Parker Lukens (October 6, 1848 – July 1, 1918) was an American conservationist, real estate investor, civic leader, and forester who believed that burned over mountains could again be covered in timber which would protect watersheds. Lukens collected pine cones and seeds of different types and conducted experimental plantings on the mountain slopes above Pasadena, California. His perseverance earned him the name "Father of Forestry." [1]

Lukens established Henninger Flats tree nursery, which provided seed stock for an estimated 70,000 trees.[2] He worked for the United States Forest Service and was acting supervisor of the San Gabriel Timberland Reserve and the San Bernardino Forest Reserve in 1906.[3]

Lukens served two terms as mayor of Pasadena and was active in municipal and civic affairs of early-day Pasadena.[4] Lukens remained prominent in civic and conservation issues until his death in 1918.[5]

23.1   Reforestation

Lukens was interested in growing plants, even before moving out to Southern California from Illinois, where he had owned and operated a nursery in Whiteside County, Illinois. By 1882 the Lukens family established a home in Pasadena. Lukens already knew of the hardwoods in his native Midwest but now the former nurseryman sought to learn about the native and non-native trees of Southern California. Among them: live oak, pepper, camphor, umbrella, eucalyptus and various citrus trees.[6]

Lukens undertook several expeditions to study the San Gabriel and San Bernardino Mountains from 1897 to 1899. He learned the "paradise" of Southern California had some serious problems as well. The increased use and misuse of resources by miners, loggers and livestock owners had devastated the lands. Wildfires caused the worst damage due to the Mediterranean climate of long, hot and dry summers which turned fires into infernos, leaving behind burned and bare hillsides, resulting in erosion and flooding during the

View of Pasadena.
Pasadena, California, Illustrated and described, by T. P. Lukens, 1886.

Cover of Pasadena, California, Illustrated and Described

rainy season.

Among the different species of trees Lukens studied, the knobcone pine was the best choice he believed, for its fire-resistant properties. The cones, which are embedded in the tree, only open and release seeds after a wildfire. He learned how to open the cones by boiling them and the method of watering and care that produced seedlings. The cone is sealed by a glaze-like resin and only opens after melting from heat of at least 200 °F (93 °C). The knobcone is well suited for reforestation as it grows on rocky hillsides in serpentine or granitic soil. Botanist Willis Linn Jepson observed that the knobcone grows on sites that are the "most hopelessly inhospitable in the California mountains".[7]

Lukens' belief in the solution of tree planting was shown by lectures he gave, as well as writings and photographs he prepared. His proposals gained support. In 1899, William Kerckhoff, president of the Forest and Water Association, paid for $50 worth of seed to give University of Southern California forestry students for planting, and in a two-week period, more than 60,000 seeds were sown by the young foresters. The next year Lukens, with support by the Forest and Water Association, planted several thousand knobcone and ponderosa pine seeds in the San Gabriel mountains above Altadena, California. Lukens' reported to the association that "ridges and crowns of hills were selected, that when the trees came into fruiting the seed would be cast in different directions down steep slopes." *[8] On a wider scale, the conservation movement was gaining momentum throughout California. In 1899, 24 organizations met in San Francisco and formed The California Society for Conserving Water and Protecting Forests. Another group formed was The Forest and Water Society of Southern California, composed of the Los Angeles Chamber of Commerce and the Southern California Academy of Science.*[9]

23.2 Tree nursery at Henninger Flats

In 1903, Lukens expanded the tree-planting enterprise with a lease on the Henninger property for the US Forest Service, of which he was an employee. Chief Forester Gifford Pinchot approved the lease in October 1903 for the clearing of 5 acres (20,000 m^2). Reforestation became official policy. Improvements included a 48' by 60' lath house and rabbit-proof fence. The first firebreak constructed in the San Gabriel Reserve was around Henninger Flats to protect the site.*[10] Lukens worked to make Henninger Flats a high elevation tree nursery that would produce seedlings for reforestation and watershed restoration efforts.

Lukens and his assistants grew more than 60,000 experimental tree seedlings at the nursery. Most of the 1,000 trees that were planted in the San Gabriel and San Bernardino reserves for the Forest Service were grown at the nursery during 1903-1907. Locally, the nursery provided 17,000 seedlings for Los Angeles' Griffith Park,*[10] the second-largest city park in California.*[11] The nursery received many orders for seed and seedlings from foresters worldwide, including Chile and Australia. Galen Clark, former Guardian of the Yosemite Grant, approved massive tree plantings, and declared this a "grand enterprise..." in a 1904 letter to Lukens.*[12]

In 1907, John Muir visited the tree nursery and was greatly impressed by the work done at the site. *A Los Angeles Times* article in June, 1912 compared Lukens to the famed Johnny Appleseed.*[13] Locally, his work at Henninger Flats was recognized when the Los Angeles County Board of Supervisors and George H. Maxwell, executive director of the National Reclamation Association, inspected the tree nursery and nearby slopes, accompanied by Lukens. Afterwards, the Pasadena *Star*, on June 21, 1915 reported-

> "It was truthfully and justly the proudest moment of Mr. Lukens' life. His work was ranked by the speakers as among the most important for the future of Southern California and as a climax Mr. Maxwell said that in his years of travel

Henninger Flats entrance

and investigation of reclamation projects he had found none more of importance to mankind than the thing Mr. Lukens had done in solving the problem of reforestation of denuded watershed areas."

> *[14]

Although Lukens' work was recognized by fellow Southern Californians, the Forest Service had a different view. The Henninger Flats nursery was closed by the Forest Service in 1908 and moved to Lytle Creek. In a 'doleful' report written in 1912, the tree planting efforts on the national forest lands were deemed a failure by Assistant Chief of Silviculture T.D. Woodbury. The Forest Service's determined efforts to convert indigenous chaparral into forests fell short due to rabbits, rodents and a baking-hot sun and the project was eventually abandoned. Woodbury's report also suggested that tree planting become a research subject and recommended it be assigned to the newly established Feather River Experiment Station, located in the Plumas National Forest.*[15]

Lukens had chosen the property for good reason-he had visited the site back in 1892 along with R.J. Busch, a Los Angeles businessman, and with the owner 's permission, they started the first experimental reforestation in California on the property by planting some selected conifers there. Those selected conifers were thriving when Lukens made a return visit about ten years later.

23.2.1 History of property

The landowner was businessman Peter Stiel who acquired the parcel through the Homestead Act. In August, 1893, Steil then sold it to his friend William Henninger who had been living on the property since 1884.*[10]

Henninger was born in Virginia in July 1817 and was among

the pioneers arriving in California in the 1800s. He was instrumental in the discovery of the first major gold strike is the San Gabriel Mountains and settled in the small hanging valley above Altadena, California. From 1884 until 1891, Henninger was the primary occupant of the 120-acre (0.49 km²) site. Henninger called the flats Clara Basin in honor of a grandchild, but this name died with him. He developed the site by building a house and a cistern for water storage. After clearing the chaparral he planted hay, melons, vegetables, and fruit and nut trees, carrying the produce to town down the steep mile and a half trail he built. He claimed water rights in 1886 for domestic and irrigation use from the first canyon north of the flats.

Henninger died in 1894, and his daughters inherited the property. The property was then sold in February 1895, by auction, to Harry C. and Harriet M. Allen of Pasadena. Selling price was $2,600. In October 1895, the Allens sold the property for $5,000 to four men (W. Morgan, J. Vandevort, J. Holmes and W. Staats). These four men then sold the property in December 1895 to the Mt. Wilson Toll Road Company for $76,600.[10] The price of this parcel had increased by 2800% in a single year. The Southern California real estate boom evidently included the mountainsides.

The property was mostly unoccupied until Lukens' second visit in 1902.

The Mt. Wilson Toll Road Company continued ownership of Henninger Flats until it was purchased by Los Angeles County in 1928. Currently, the county fire department manages the site, which is now known as the Henninger Flats Conservation Center and operates a museum, campground, and tree nursery.[10]

23.3 Forester

Lukens worked for the US Forest Service from 1900 to 1906. His first association with the agency was through Muir, who recommended him to Chief Forester Gifford Pinchot in 1899. Next, State Senator Delos Arnold, a prominent Pasadenan, recommended Lukens and by May 1, 1900, Lukens was offered the honorary position of Collaborater at $300 a year. His duties included investigative trips and he wrote detailed reports on other forests, as well as attended meetings and conventions. In 1906 he was promoted to the position of Acting Supervisor and was in charge of the San Bernardino Forest Reserve and the San Gabriel Timberland Reserve. The San Gabriel Reserve was the first federal reserve established in the state.[16]

Lukens knew how serious the threat of wildfire was in the mountains and pushed for more firefighters, which the Forest Service granted. A joint survey by the state of California and the US Forest Service (then Bureau of Forestry) found the forest conditions in deplorable shape. Watersheds were

being damaged yearly by fire, overgrazing, and land clearing. Logging had badly deteriorated the forests. The survey results produced several actions, one of which was the passage by the California legislature of the Forest Protection Act that created a State Board of Forestry with a State Forester.

Lukens hired 55 men for firefighting and other duties, but less than one year later, he was instructed to reduce the force to 25 which he found unacceptable.[3] He wrote letters to Pinchot and influential friends in an attempt to bring political pressure to rescind the order.[3] To complicate matters, he had requested and been granted a three-month personal leave of absence.[3]

While Lukens was on vacation in Yosemite, Chief Pinchot transferred him from supervisor and into a "special position" with forestry extension services.[17] The agency also made expense and payroll mistakes which cost him money, He tendered his 11-word resignation on August 12, 1906[17] which also ended his work at Henninger Flats.

23.4 Conservationist

Lukens made several trips to Yosemite National Park, each time his interest in every aspect of nature intensified, and led to research, correspondence and interviews with experts for whom he collected various specimens and photographed specific items. Lukens joined the fledgling Sierra Club in 1894 after a visit to Yosemite and the Sierra Nevada Mountains. A self-taught photographer, he prepared albums of 100 to 150 photos of trees and mountains for friends such as Senator John Bard and Alice Eastwood, botanist at the California Academy of Sciences, among others.[18]

One particular trip in 1895 to Yosemite's Hetch Hetchy Valley was for the sole purpose of meeting the famed conservationist John Muir, who was reported to be roaming the valley. Lukens outfitted himself at Crocker's Station, a stagecoach stop and resort, with food, two pack animals and Elwell the guide. A few miles south of the valley, he came across a man wearing a rumpled suit, vest, white shirt; without food, pack animals or companions.

He was indeed Muir, and Lukens persuaded him to join his well-stocked entourage and they headed back to Hetch Hetchy Valley, but only after Lukens took a photo to memorialize the event. The photogragh is titled "John Muir Resting".[19] The two men discussed topics on forest preservation, as well as watershed restoration and Southern California wildfires.[19] Muir was familiar with Southern California, having visited and climbed there in 1887.[19]

Lukens and Muir discovered that they shared a friend: Jeanne Smith Carr. Her husband, Dr. Ezra Carr had been

Muir, photographed by T.P. Lukens, 1895

Muir's professor in Natural Science at the State University of Wisconsin. The Carrs had moved to Pasadena in the 1880s. Lukens was on several Pasadena committees with Jeanne Carr and shared an interest in tree-growing. As the two mountaineers, Muir and Lukens, explored the area, Lukens studied pine trees, soil types and growth habits, noticed height and age, and as always, collected pine cones and absorbed Muir's essays on conifers.

Muir with Lukens at Crocker's Station, 1895

The sugar pine was Muir's favorite, which he described as "surpassing all others, not merely in size but in lordly beauty and majesty." The sugar pine is the largest and tallest pine of the pine trees[20] with recorded heights exceeding 200 feet (61 m), and has the longest cones, up to 24 inches (610 mm) in length.[21] During the last week of the trip, they camped at Tenaya Lake, where they met Gertrude Towne and Nellie Anderson and their mule, Plumduff. Muir and Lukens led the intrepid adventuresses up Mt. Conness to the summit at 12,566 feet (3,830 m) where Miss Towne drew sketches of the distinctive peak. Mount Dana, elevation 13,052 feet (3,978 m), was climbed the next day by the group before Muir had to return home to Martinez. Muir sent Lukens a copy of *Picturesque California*, volume II with a note to read Jeanne Carr's chapter on Los Angeles County.[22][23] Lukens and Muir maintained a friendship until Muir's death in 1914.

23.4.1   Sierra Club

At Muir's urging, Lukens became actively involved in several Sierra Club campaigns: to purchase privately held toll roads for public use into Yosemite National Park, and to protect Hetch Hetchy Valley in Yosemite.

John Muir, now president of the Sierra Club, appointed Lukens to the Club committee of Publications & Communications and directed him to write articles and speak about government-owned toll roads and of California relinquishing both Yosemite Valley and the Mariposa Grove of Big Trees to the federal government, which were then under state management. In a letter to Lukens, Muir urged Lukens to "stir up the Senate Interior Section and make their lives wretched until they do what is right for the woods." [24]

Lukens corresponded with various people and local newspapers concerning the Sierra Club campaign. In a letter to the Pasadena *Daily News*, Lukens wrote- "The small portion of this great park now owned and controlled by the State of California should be receded to the national government, and have it all under one management. then there would be no need of petition for roads. The whole park would be cared for as the Yellowstone now is..." -Pasadena *Daily News*,undated.[25]

Lukens spent most of 1896 actively involved in the politics of conservation, except for three months spent mountaineering in Yosemite, which ended in controversy a year later. He was accompanied by Walter Richardson, a young Pasadenian. Both Lukens and Richardson wrote about the horseback trip for the Sierra Club *Bulletin* and the weekly Pasadena newspaper *Town Talk*. Highlights of the expedition included trail-building into steep Tehipite Valley, which Richardson named "Lukens Trail" . (A young hiker by the name of Amy Racina would write a non-fiction book of her 60-foot (18 m) fall, survival and rescue in this same valley in 2003.) They had special permission to carry firearms for "personal protection" only, as no hunting was allowed in the park. While the men stayed at a private cattlemen's cabin, five bears were killed.*[26] No guilty conscience appears in Lukens' journals, possibly because he was on private land. The United States Army was the park's law enforcement and viewed the situation as a violation of the permit. The reports and the permit were sent to the Sierra Club in January, 1897. The incident led to an exchange of letters between Lukens and Muir. Lukens was unapologetic about the hunting incident but Muir in turn advocated the creation of wildlife refuges within the national park and forest reservations. "Sort of a wild beast paradise," Muir described it in a letter to Lukens.*[27] Both Muir and Lukens continued their conservation efforts.

During 1897 Both Lukens and Muir were active in different areas of the evolving conservation movement. "I've been writing about the forests, doing what little I can to save them," Muir wrote to a colleague, Henry Fairfield Osborne. Lukens that year began his campaign for watershed protection of the San Gabriel Reserve. He advocated fire protection, tree planting and removal of stockmen from the reserve. He and others undertook two pack trips into the reserve in 1897 and one expedition in 1998. Lukens wrote little, but was a prolific photographer, documenting trees, rocks and conditions during the pack trips. Lukens was often quoted in newspapers, because he gave talks to community groups about his ideas for preserving the watersheds. In an 1898 article in the *Los Angeles Times*, Lukens was quoted terming the driving of stock in the mountains as "Hoofed Locusts." His argument (and Muir's) was that sheep destroyed the mountain meadows by grazing away all vegetation and that stockman were guilty of setting wildfires at the end of the season to create more meadow for the following year.

Lukens's first tree planting expedition on the mountain ridges occurred in 1900. Fifteen years later the Los Angeles County Board of Supervisors ascended Mt. Wilson to view the earlier work: "As the father of reforestation of these mountains, it was a happy day for T.P. Lukens who went with the party and looked with a feeling of pride on the result of his efforts many years ago," The *Los Angeles Times* reported June 12, 1915. Although Lukens' advocacy of reforestation was well received by local officials, it does not appear that further reforestation efforts were carried out. The 1915 article went on to predict, "as a result of this foresight and the trip yesterday, 1,000,000 trees perhaps will be planted on the barren and burned stretches of the mountains this year." No news report exists indicating the plan was carried out. Lukens was then four months shy of his 67th birthday.

23.5 Early life

Lukens was born into a German Quaker family in Ohio on October 6, 1848. When Lukens was 6 years old, the family moved to Illinois and began a nursery business. He enlisted in the US Cavalry at 20 and two years later received an honorable discharge.*[28] After military service, he married his first wife, Charlotte Dyer. He began his own nursery business in Whiteside County, Illinois growing fruit and ornamental trees. The nursery also included a floral department. He involved himself in local community affairs, including the job of tax collector from 1873 to 1876.*[28] Their only child, Helen was born January 9, 1872 and when she turned 8, he moved the family to California to improve a deteriorating financial and health situation.

23.6 Pasadena settlement

The Lukens family settled in Pasadena in December, 1880. Theodore Lukens was active in his new community and by 1884, he was elected Justice of the Peace, as well as a member of the new Republican Committee. Two years later, the Southern California land boom swept Lukens, the first real estate agent in Pasadena, into a wealthy position. He is credited with selling his large interests in the Raymond Tract, one of the earliest subdivisions of acreage at the time. The year 1886 was a busy one, the town of Pasadena incorporated on June 14, the Pasadena National Bank began accepting deposits and the 201-room Raymond Hotel opened in November for snow-weary guests. Lukens' was able to semi-retire from his real estate business, he sold his properties and traveled. He wrote the first advertising booklet for the town, in the hope of getting the hotel guests to settle in Pasadena. Rich in superlatives, the booklet is titled *Pasadena, California, Illustrated and Described* and is now a collector's piece.

In April, 1888, the Board of Trade was organized with Lukens as a charter member. The Board was an early-day chamber of commerce for Pasadena, to attract industry and promote the city.

Pasadena's public library opened in 1890, with support of Lukens in fundraiser events such as the 10-day Art Loan Exhibit in February, 1889. Each day had a different historical theme with speeches by well-known people. On Russian Day Lukens exhibited his Alaskan artifacts. On Forestry Day, the speakers were Abbot Kinney, the State Forestry Commissioner and Jessie Benton Fremont, wife of John Fremont. The new library's dedication ceremony included Lukens, Kinney and Amos G. Throop, a Pasadena Trustee (city councilmember).

In 1891, Lukens was a cashier for the Pasadena National

Bank and by 1895, bank president.[*][29] He helped the local banking industry during the Panic of 1893 by enlisting aid from bankers in Los Angeles. Although short-lived, the Panic of 1893 was severe with over 150 national and 172 state bank failures.[*][30]

23.6.1 Mayor

National Bank of Pasadena, 1895

Lukens served as president (mayor) of Pasadena twice; the first two-year term was 1890–1892 and a partial term from 1894 to 1895 when he was replaced by John S. Cox.[*][4] Lukens's second term as Mayor ended on January 2, 1895 as he was summarily "retired" from office because he refused to endorse a resolution allowing a franchise by Southern Pacific Railroad for a railway through Arroyo Parkway to Broadway Street to Green Street. There was already a Santa Fe line, the electric Mount Lowe Railway to Echo Mountain and a local rail line to Los Angeles. The council voted against their president and for the franchise and Lukens left with his convictions intact. Public opinion was expressed in newspapers and favored Lukens' decision.[*][31] "To be disposed from a position for no other fault than standing for principles, is an honor rather than a disgrace", commented the Los Angeles Daily *Hotel Gazette*.[*][31] During Lukens' first term as mayor, he had the pleasure of receiving two dignitaries: President Benjamin Harrison and First Lady Caroline, who visited Pasadena on March 23 and 24, 1891 during their tour of the west.[*][32]

Lukens was a member of many local boards and committees, such as the Pasadena World's Fair Committee and provided orange trees and 60 palm trees for the exhibit in the World's Columbian Exposition held in Chicago, Illinois in 1892 that commemorated the 400th anniversary of America's discovery by Christopher Columbus. He was awarded a medal of honor with his name inscribed. In addition,

Lukens served as president of Pasadena Mutual Building and Loan Association, board member of two schools which were the Los Angeles State Normal School (college) and California Institute of Technology, and he was a member of the Los Angeles Chamber of Commerce.

23.7 Later years

Lukens became a grandfather in 1891 to Charlotte (or "Lottie") Jones, and in 1893, Ralph Jones. The health of Lukens and his wife Charlotte was declining by 1903, and Charlotte died on December, 1905. Lukens then requested his three-month leave of absence from the Forest Service. Lukens remarried July, 1906 to Hannah Sybil Swett, a long-time family friend and Christian Scientist. He subsequently adopted his wife's religion.

In his waning years, Lukens remained active in local affairs and conservation. He promoted the establishment of a park above Devil's Gate Dam, which eventually became a reality as Oak Grove Park. Two months before his death, Lukens wrote a report on conservation and forestry for the annual meeting of the Bureau of Forestry.

Theodore Lukens died July 1, 1918 and is buried next to his first wife at Mountain View Cemetery in Altadena, California. H. Sybil Swett Lukens died in December of that same year.[*][33] The *Los Angeles Times* published an obituary on Theodore Lukens on July 5, 1918, which stated in part, "There is a new, growing forest of neglected pine trees on the slopes of the Sierras [San Gabriel Mountains] above Pasadena that stand as a living proof that Lukens could have reforested our mountains if we had only given him the help he asked. He would have restored to us the lost beauty of the hills and he would have, at the same time, saved us from a disaster which may one day overtake us." [*][34]

A Lukens Memorial Forestry Society was begun by poet John S. McGroarty and Marshall V. Hartranft, co-founder of the socialist Little Lands Colony. The Society fostered reforestation and endured until the 1950s.[*][33]

23.7.1 Helen Lukens

Helen became a published writer and photographer of some note during the first decade of the 20th century. The photograph of John Muir on the introductory page of Linnie Marsh Wolf's Pulitzer prize-winning biography of Muir was taken by Helen Lukens.

She was married in 1890 to Edward Everett Jones, the Bank of Pasadena cashier who replaced Lukens, and they had two children, Charlotte ("Lottie") and Ralph. Lottie became a favorite of John Muir and she considered him a second father. The marriage to Edward Jones did not last, and in 1906, Helen married James H. Gaut.

Helen Lukens Jones was the first woman to travel by automobile up the Mount Wilson Toll Road Company's nine-mile (14 km)-long, narrow, 10% grade road that went from Eaton Canyon through Henninger Flats and then to the summit.*[35]

Decades after her father's death, Helen sold his records, diaries and papers to the Huntington Library which are now used in various studies. The collection is over 3,600 pieces.*[36]

23.8 Tributes

Lukens' home on N. El Molino Avenue in Pasadena, California

The "Father of Forestry" Theodore Lukens is memorialized today with a mountain, a lake in Yosemite, and his home listed on the National Register of Historic Places.

Sister Elsie Peak was renamed to Mount Lukens by the Forest Service in the 1920s. Mount Lukens is the highest point within the city limits of Los Angeles at an elevation of 5,066 feet (1,544 m).*[37]*[38]

Robert Bradford Marshall, a 30-year veteran of the US Geological Survey named a lake in Yosemite National Park for Lukens in 1894.*[39]

The Lukens family home on 267 N. El Molino Avenue in Pasadena was added to the National Register of Historic Places on March 29, 1984.*[40] The house was designed by Harry Ridgeway, who also designed the city jail and some of Pasadena's early Methodist and Episcopal churches.*[41]

23.9 Notes

[1] Godfrey, Anthony pp. 34-35

[2] Angeles National Forest Fire Lookout Association Newsletter, p.4

[3] Sargent, Shirley p.66

[4] Wood, John Windell p. 199

[5] Sierra Club's John Muir Exhibit.

[6] Sargent, Shirley p. 8

[7] Lanner, Ronald p.97

[8] The *Los Angeles Times*, "Creating New Forests", Feb.9, 1900.

[9] Godfry, Anthony pp.48-9

[10] Los Angeles County Fire Department

[11] *The 100 Largest City Parks*. from The Trust for Public Land.

[12] Sargent, Shirley p.53

[13] Sargent, Shirley p.55

[14] Sargent, Shirley p.83

[15] Godfrey, Anthony p.95

[16] Godfrey, Anthony p.38

[17] Sargent, Shirley p.67

[18] Sargent, Shirley p.24

[19] Sargent, Shirley p.27

[20] Stuart, John D. and John Sawyer p.78

[21] Lanner, Ronald p.11-12

[22] Sargent, Shirley p.33

[23] Yosemite Online Library, John Muir writings.

[24] Sargent, Shirley p.45

[25] Sargent, Shirley pp 34-35

[26] Sargent, Shirley pp 40-41

[27] Sargent, Shirley p 44

[28] Sargent, Shirley p. 2

[29] Biographer Shirley Sargent puts the date at 1893

[30] Northwestern Univ. Panic of 1893 webpage.

[31] Sargent, Shirley p. 19

[32] The *New York Times*, "The Presidential Party Visit Pasadena and Santa Barbara" published April 25, 1891.

[33] Sargent, Shirley p.86

[34] *Los Angeles Times*, "The Late T.P.Lukens". Jul 5, 1918; ProQuest Historical Newspapers *Los Angeles Times* (1881 - 1986) p. II4

[35] Sargent,Shirley p.59

[36] Sargent, Shirley p. 23

[37] Robinson, John W. p.47

[38] USGS Query results.

[39] *California Place Names...*, authors Erwin Gudde, Wm. Bright p.220

[40] National Park Service, National Register Information System search results for "Lukens"

[41] *Los Angeles Times.* Los Angeles, Calif.: Apr 11, 1990. p. 5.B

23.10   References

- Angeles National Forest Fire Lookout Association Newsletter, August 2006 accessed 16 Oct. 2008

- Godfrey, Anthony *The Ever-Changing View-A History of the National Forests in California* USDA Forest Service Publishers, 2005 ISBN 1-59351-428-X

- Gudde,Erwin Google Books search results-*California Place Names...*, authors Erwin Gudde, Wm. Bright accessed 16 Oct. 2008

- Lanner, Ronald M. *Conifers of California.* Cachuma Press, Los Olivos, CA 1999 ISBN 978-0-9628505-4-7

- Los Angeles County FD History webpage accessed 16 Oct. 2008

- Robinson, John W. *Trails of the Angeles* Wilderness Press, sixth ed. 1990

- Sargent, Shirley *Theodore Parker Lukens-Father of Forestry* Dawson's Books (Los Angeles, CA) 1969 *no ISBN

- Sierra Club's John Muir Exhibit accessed 16 Oct.2008

- Stuart, John D., John Sawyer, *Trees and Shrubs if California* Univ. of California Press 2001

- Wood, John Wendell *Pasadena, California, Historical and Personal...*Published by The author, 1917

23.11   External links

- Online Archives of California, **Theodore Lukens collection**.

- Sierra Club, Angeles Chapter Bulletin: Essay on the history of Sister Elsie Peak, now named Mount Lukens.

- Blog post about Helen Lukens

Chapter 24

Million Tree Initiative

The **Million Tree Initiative** refers to the on-going environmental projects that multiple cities have individually committed to, aimed at increasing the urban forest through the planting of one million trees. Cities that are known to be currently involved in this initiative are: Los Angeles, New York City, Shanghai, Denver and London, Ontario. A common motive shared between these participating cities is, according to their mission statements, the reduction of carbon dioxide in the air to reduce the effects of global warming.

24.1   History

In May 2006, Mayor Antonio Villaraigosa made Million Trees LA one of his campaign promises. The Los Angeles project is founded by a mix of federal money and municipal funding, charities, and corporate donations. It was one of among forty winners from 200 nominees to obtain a United States Environmental Protection Agency (EPA) Environmental Award in 2009.*[1]

The Mile High Million, an initiative started by then Mayor John Hickenlooper, is a similar program in Denver, Colorado. This was announced by Hickenlooper in his 2006 State of the City Address.*[2]

On April 22, 2007, Mayor Michael Bloomberg revealed goals of planting one million trees by 2017 as part of PlaNYC, a plan designed for the sustainability of New York City.*[3] In the same year, China began its own tree planting program for Shanghai, with the same goal for one million trees.

The million trees program began in London in 2011.

24.2   Benefits

Main article: Urban forest § Benefits

24.3   References

[1] EPA. 2009 Environmental Awards Retrieved September 27, 2011.

[2] Hickenlooper, John (July 12, 2006). State of the City Address 2006. Retrieved September 27, 2011.

[3] The City of New York (April 22, 2007). MAYOR BLOOMBERG PRESENTS PLANYC: A GREENER, GREATER NEW YORK Retrieved February 28, 2011.

24.4   External links

- Million Trees LA
- Million Trees Denver
- Million Trees NYC
- Million Trees Shanghai
- Million Trees - London, Ontario

Chapter 25

Monterey County reforestation

Monterey County in California, U.S., has a commitment to its environment, and more specifically to its forestation, and offsetting the community's carbon footprint. To that end, a number of resources have been committed to preserving and improving this unique ecosystem and its reforestation.

LMP Class of 2008 Reforestation Project

25.1 Pine Forests on the Monterey Peninsula

A patchwork pine forests grow across the Monterey Peninsula in California, creating a rich green habitat that delights residents and visitors alike. Although the locally native Monterey Pine - *Pinus radiata* tree species is widely grown in landscapes and on plantations in the Western Hemisphere of Northern America, the native Monterey Pine ecosystem is one of the rarest forest ecosystems in the world. Only a few thousand acres of the endemic trees exist in four locations along the Pacific Ocean on the Central Coast of California. The Monterey Peninsula is home to the largest of these stands, but the trees there are threatened by impacts from development, non-native invasive species, and disease.*[1]

The local Monterey pine forest provides numerous benefits to the region's economy, from its intrinsic beauty that attracts tourists, to recreational settings for residents and visitors, to valuable ecological services, such as watershed protection and enhanced air quality. However, except for a few small sites and the Point Lobos State Reserve, much of the remaining forest is not protected within conservation areas. According to The Monterey Pine Forest Watch, "Half of our native forest has already been removed" and "Much of the remaining forest is in private hands and subject to development." *[2]

25.1.1 Urban Forestry

Even sites that are protected from tree removal, such as Veteran's Memorial Park, have been impacted by human activities and tree diseases, such as pine pitch canker. According to Robert Reid, the head of the City of Monterey's Urban Forestry program, "Since the disease was first discovered in 1986, *hundreds of pines of all ages have died* and their loss has had a noticeable impact on the forested areas and landscapes of the Peninsula." *[3] In the Spring 2007 issue of City Focus, he describes how scientists have been unable to find a cure or stop the spread of the pitch canker fungus: "The infection causes pines to ooze sap, which results in greater susceptibility to attack from destructive pine bark beetles." The fungus is spread as beetles move from tree to tree.

The local city forestry departments have small staffs and are charged with the responsibility of caring for large areas of public lands. In addition, city arborists are called upon to advise on property issues, construction impacts and risk assessment. The City of Monterey itself maintains more than 19,000 trees, in parks and along streets, and about 300 acres (1.2 km²) of Monterey Pine forest. Monterey currently spends nearly $1 million annually on its urban forestry program, or about $33 per citizen. Carmel spends about $450,000, which with its smaller population translates to about $112 per citizen. An active volunteer group, Friends of Carmel Forest, supports Carmel's tree planting, surveying and education activities. Carmel's City Council also recently voted to approve a contract with a renowned tree care specialist, Barry Coates, to provide a study of Carmel's forest. The Council also approved spending up to $50,000 to implement changes recommended by the forest study.*[1]

In the City of Monterey, crews trim an average of more than 1,800 trees and remove 150 annually. In 2005, a total of 338 trees were planted or replaced on City property, and 400 native tree seedlings were donated for planting on private property. In addition, the City of Monterey Urban Forestry Section provides tree maintenance services to the Presidio and Naval Postgraduate School. Given the tremendous workload and the importance of the local forests, volunteer groups such as Leadership Monterey Peninsula are critical partners in maintaining and improving the region's quality of life.*[1]

25.2   Individual City Information

Polled in 2007/8 by the Leadership Monterey Peninsula Reforestation Group, the following is a breakdown of the cities in Monterey County, and their reforestation efforts.

25.2.1   City of Monterey

Current Forestation Activities:
As of 2008, the City of Monterey has 9 employees in the Urban Forestry Department. The City of Monterey works with various volunteer groups to plant trees in high-needs areas; there is also a Neighborhood Improvement Program (NIP) in which residents can request that trees be planted in certain residential areas, such as along the side walks. Their budget for tree-planting and maintenance in your city was $1 million, including payroll for 9 staff in 2008. This funding comes from taxes and periodic grants; maintenance through the urban forestry dept. and volunteers. Historical efforts for this department included spring 2003 stats: 768 on city property, 1500 seedlings donated for planting on private property; average number of trees is 300-500 annu-

ally. Other that have assisted in reforestation efforts include Leadership Monterey Peninsula, Boy Scouts of America, and schools that have periodically assisted with tree planting projects, as well as Volunteer Gardener Program – this group tends to do more ornamentals and flowers for the city and Monterey Green Action (Yahoo groups) – has volunteers available.

25.2.2   City of Carmel

Current Forestation Activities:
Carmel's Department of Forest and Beach is responsible for the maintenance of and improvements to the urban forest. The majority of their time is spent pruning, planting, watering, treating insects, and removing dead trees. The city's annual budget for tree-planting and maintenance is $458,000 including beach maintenance and all Forest and Beach Departmental services. This department, encompassing three employees, takes care of all the arboricultural needs of over 13,000 trees growing on public property.

25.2.3   City of Marina

Current Forestation Activities:
Marina has been named a "Tree City USA" community by the Arbor Day Foundation.
Had to meet four standards to get this designation:
1) has a tree board or department
2) has a community tree ordinance
3) has a community forestry program w/ an annual budget of at least $2 per capita (budget is $50,554.36; population is 18,824)
4) has an Arbor Day Observance and Proclamation (Dec. 17, 2007)

The tree committee has recently been reorganized; it is now a subcommittee of the planning commission. Tree planting activities are now done by a non-profit: "Marina Tree and Garden Club". Their mission includes "Improve the streets and public areas of the city of Marina by planting and cultivating trees and gardens therein"

The city does not have an arborist or a tree-planting program.

25.2.4   City of Seaside

Current Forestation Activities:
No formal reforestation program. Parks and Recreation Dept has an annual budget of approximately $150,000, and did plant 50+ trees in 2007. A limited number of trees are taken care (watered) of by the city parks department.

There is an established relationship with their neighborhood associations to assist with planting of trees within Seaside.

25.2.5   City of Del Rey Oaks

Current Forestation Activities: Currently looking for Oak tree donations

25.2.6   City of Pacific Grove

Current Forestation Activities:
Pacific Grove's tree planting program is called "Trees for Pacific Grove", a public private partnership with this non-profit. In 2007, Trees for PG raised over $10,000 through sponsorships. Individuals can become a Seedling Sponsor for $75 or a Sapling Sponsor for $250. During the same time period, Pacific Grove also planted over 1,000 trees and gave away another 500 native trees to local residents to plant on their own property.
Sustainable Pacific Grove is another organization interested in tree planting in PG.

25.2.7   City of Sand City

25.2.8   City of Salinas

25.3   Notes

[1] Leadership Monterey Peninsula Class Projects, "Small-Scale Reforestation in an Urban Forest"

[2] "Our Monterey Pine Forest: A Forest Under Siege," pg. 3, The Monterey Pine Forest Watch.

[3] "Monterey Pines Survival," Robert Reid, City Focus, Volume XXII, Number 1, Spring 2007, p. 5.

25.4   See also

- Reforestation

- Monterey, California

- Monterey Bay

- Monterey Peninsula

- Category: Forestry and sustainability

- Sustainable forest management

25.5   Bibliographies

- Plants.   Research conducted in Northern Santa Lucia Mountains, Big Sur, and surrounding areas, 1994-1997.   Santa Lucia Natural History Symposium (sponsored by Esalen Institute and University of California Big Creek Reserve. http://www.redshift.com/~{}bigcreek/projects/ natural_history/slnhs_bibliography/slm_plants.html

- Selected biological, ecological, and genetic literature relevant to Monterey pine (360 references in alphabetical order by author) / Genetic Resources Conservation Program, University of California http://www. grcp.ucdavis.edu/projects/MPRefs/MPRefList.htm

- Technical bibliography on Monterey Pine, Pinus radiata (updated 23 October 1998) / Galen Rathbun, Piedras Blancas Field Station, Western Ecological Research Center, US Geological Survey, San Simeon, CA 93452. http://www.greenspacecambria. org/Documents/MontereyPinesBibliography.pdf

25.6   References

- The Monterey Pine through geologic time / by Frank Perry, Research Associate, Santa Cruz Museum of Natural History (reproduced from the Monterey Bay Paleontological Society Bulletin, July–September 2004. © 2004 Frank Perry) http://evolution.berkeley. edu/evolibrary/article/0_0_0/montereypines_01

Chapter 26

Mount Airy Forest

The **Mount Airy Forest**, in Cincinnati, Ohio, was established in 1911. It was one of the earliest, if not the first, urban reforestation project in the United States. With nearly 1,500 acres (6.1 km^2), it's the largest park in Cincinnati's park system.[3]

26.1 History

The originally forested land was cleared for agricultural use in the 19th century, but years of poor grazing and agricultural practices led to severe erosion and poor soil composition. As quoted in a 1914 *Cincinnati Times-Star* editorial, a farmer facetiously remarked that his farm (in Westwood) "was a good one when he first took it up but that since he had cleared off all the trees it had slid down the creek and was to be found somewhere in the neighborhood of New Orleans." [4]

According to the National Park Service:

> Established in 1911, the Mount Airy Forest covers an impressive 1459 acres and includes natural areas, planned landscapes, buildings, structures, and landscape features. The numerous hiking trails, bridle paths, walls, gardens, pedestrian bridges, and various other improvements within Mount Airy Forest reflect the ambitious park planning and development that took place in Cincinnati in the early-to-mid-20th century. Conceived as the nation's first urban reforestation project, the park has developed over the years—especially during the Depression and post-World War II period- into a park with a variety of areas, spaces and structures designed to accommodate recreational, social, and educational activities. Today it continues to offer a large expanse of protected land within the city limits where the public can enjoy the richness and diversity of nature.[5]

In the largest reforestation program undertaken by a city seen until that time, the barren land was restored to a park largely in the 1930s by the Civilian Conservation Corps (CCC).[6] The rustic CCC structures are still standing and are listed on the National Register of Historic Places.[1] The park now includes 700 acres of reforested hardwoods, 200 acres of forested evergreens, 269 acres of wetlands, 170 acres of meadows, and a 120-acre arboretum.[7]

The park was listed on the U.S. National Register of Historic Places on April 13, 2010.[2] The listing was announced as the featured listing in the National Park Service's weekly list of April 23, 2010.[8]

Gazebo in Mt. Airy Forest Arboretum

26.2 Amenities

The park offers hiking trails, an 18-hole disc golf course,[9] and a 2-acre (0.81 ha) dog park.[10]

26.3 References

[1] Nancy Recchie (May 2008). "National Register of Historic Places Registration: Mount Airy Forest" (PDF). National Park Service. Retrieved 2010-05-11. (72 pages, with maps and historic and modern photos)

[2] "Announcements and actions on properties for the National Register of Historic Places for April 23, 2010". *Weekly Listings*. National Park Service. April 23, 2010. Retrieved 2010-05-11.

[3] Smith, Steve; et al. (2007). "Major Parks". *Cincinnati USA City Guide*. Cincinnati Magazine. p. 19. Retrieved 2013-05-06.

[4] "Mt. Airy Forest". Mount Airy Town Council.

[5] "Weekly Highlight 04/23/2010 Mount Airy Forest, Hamilton County, Ohio".

[6] Pender, Linda (Jul 1983). "Pick a Park, Pack Up and Go". *Cincinnati Magazine Jul 1983*. p. 65. Retrieved 2013-05-08.

[7] "Mount Airy Forest". City Beat. Retrieved 2011-12-30.

[8] "Weekly List Actions". National Park Service. Retrieved 2010-05-11.

[9] Doane, Kathleen (May 2002). "Our Glorious Parks". *Cincinnati Magazine*. p. 52. Retrieved 2013-05-18.

[10] Burwinkel, Beth (1 January 2005). *A Bark in the Park: The 44 Best Places to Hike with Your Dog in the Cincinnati Region*. Cruden Bay Books. p. 19. ISBN 978-0-9744083-6-1.

26.4 External links

- Mt. Airy Forest - Cincinnati Parks Dept.

- Unofficial trail maps

Chapter 27

The National Forest (England)

Conkers Discovery Centre at Moira in the heart of the National Forest

Location of the national forest.

The National Forest is an environmental project in central England run by **The National Forest Company**. Areas of north Leicestershire, south Derbyshire and southeast Staffordshire, 200 square miles (520 km²) are being planted, in an attempt to blend ancient woodland with new plantings to create a new national forest. It stretches from the western outskirts of Leicester in the east to Burton upon

Trent in the west, and is planned to link the ancient forests of Needwood and Charnwood.

27.1   The National Forest Company

The National Forest Company is a not-for-profit organisation established in April 1995 as a company limited by guarantee.*[1] It is supported by the Department for Environment, Food and Rural Affairs (DEFRA), with the aim of converting one third of the land within the boundaries of the National Forest (135 km², 33,000 acres) to woodland, by encouraging landowners to alter their land use. It is described as "a forest in the making" and it is hoped to increase tourism and forestry-related jobs in the area.

Around 8 million trees have been planted, tripling the woodland cover from 6% to around 18%.

27.2   Planting

Approximately 85% of the trees planted are native broadleaf species with some of the most commonly planted species are: English oak, ash, poplar, Corsican and Scots pine.*[2]

The transformation of the landscape is beginning to take effect as the first tiny whips planted in the early 1990s are growing into substantial trees.

27.3 Attractions

At the centre of the National Forest, is *Conkers*, a visitor centre located just outside the village of Moira, Leicestershire. There is also a visitor centre with wildlife walks and playgrounds at Rosliston.

Other attractions include:

- Ashby Canal

- Ashby Canal Association

- Ashby Canal Trust

- Ashby Castle, Ashby-de-la-Zouch

- Ashby de-la-Zouch museum, Ashby de-la-Zouch

- Bardon Hill - highest point in the National Forest at 912 feet (278 m) above sea level.

- Barton Turns Marina, Barton under Needwood

- Battlefield Line Railway,

- Billa Barra Hill

- Beacon Hill, Leicestershire

- Beehive Farm, Rosliston

- Bradgate Park

- BCTV Conservation Holidays

- Calke Abbey, Ticknall

- Catton Hill

- Claymills Victorian Pumping Station

- Conkers, Moira

- Crackpotz Ceramic Cafe,

- Croxall Lakes

- Donington le Heath Manor House

- Ferrers Centre for Arts and Crafts, Ashby de-la-Zouch

- Flagship Diamond Wood

- Foremark Reservoir, Foremark

- Forest Four Wheel Drive

- Fradley Junction

- Grace Dieu Priory

- Greenwood Days

- Hill Hole Quarry

- Kedleston Hall

- Lakeside Lodge Tearooms

- Loughborough Outwoods

- Martinshaw Woods, Groby

- Melbourne Hall

- Moira Furnace

- Mount St. Bernard Abbey

- National Forest Llama Treks

- National Memorial Arboretum

- North West Leicestershire Museums

- Rosliston Forestry Centre

- Seale Wood

- Sence Valley Forest Park

- Sharpes Pottery Museum, Swadlincote

- Shortheath Fisheries

- Skylark Holidays

- Snibston Discovery Park

- Staunton Harold Reservoir & Visitor Centre

- Sudbury Hall

- Swadlincote Ski & Snowboard Centre

- Swithland Wood

- TG Green Pottery

- The Cattows Farm Shop & Heather

- The Museum of Brewing at Coors Visitor Centre

- The National Forest Maze

- The Shoulder of Mutton

- Thornton Reservoir

- Tropical Birdland, Leicestershire

- Twycross Zoo

- Uttoxeter Heritage Centre

- Uttoxeter Racecourse

- Willesley Wood

As well as Ashby de la Zouch, the towns of Burton upon Trent, Swadlincote and Coalville are also located within the forest area.

27.4   See also

- Reforestation

- Plant A Tree In '73

27.5   References

[1] The National Forest, About us

[2] The National Forest, Frequently Asked Questions.

27.6   External links

- The National Forest Visitor pages

- The National Forest Website

- Rosliston Forestry Centre

- Conkers Website

Chapter 28

NZ Native Forests Restoration Trust

Founded in 1980, the **NZ Native Forests Restoration Trust** is an organisation involved in forest restoration.

The Trust has acquired land at the rate of 250 ha per year to protect important species, restore their habitats and to improve the quality of waterways. It now has 24 reserves throughout the North Island and one in the South Island with a total of nearly 6,000 ha of protected native forests.

Sir Edmund Hillary was the foundation patron of the trust until his death in 2008. Sir Paul Reeves then became patron until his death in 2011.[*][1]

The Trust publishes the *Canopy* newsletter two or three times per year.

28.1   See also

- List of environmental organizations

28.2   References

[1] *Canopy* (Auckland: NZ Native Forests Restoration Trust) (57), Spring 2011, ISSN 1170-3172 Missing or empty |title= (help)

28.3   External links

- Native Forest Restoration Trust

Logo of the Trust.

Chapter 29

Pottiputki (tool)

Pottiputki and worker

Pottiputki is a planting tool that was created by Tapio Saarenketo in the early 1970s. It is an efficient tool for manual planting of containerized seedlings. The planters can work in an ergonomically correct position while maintaining high productivity, making the task both fast and comfortable.[1] It is more effective, but more expensive than the traditional mattock.

The tool is available in multiple sizes and tube diameters.

29.1 See also

- Hoedad

- Kidney tray (tool)

- Tree planting

- Tree planting bar

- Reforestation

29.2 References

[1] "Pottiputki". Bccab.com. Retrieved 2014-03-26.

Chapter 30

Reducing emissions from deforestation and forest degradation

"REDD" redirects here. For the UN initiative, see United Nations REDD Programme. For other uses, see Redd (disambiguation).

Reducing emissions from deforestation and forest degradation (**REDD**) is a mechanism that has been under negotiation by the United Nations Framework Convention on Climate Change (UNFCCC) since 2005, with the objective of mitigating climate change through reducing net emissions of greenhouse gases through enhanced forest management in developing countries.

In the last two decades, various studies estimate that land use change, including deforestation and forest degradation, accounts for 17-29% of global greenhouse gas emissions.[*][1][*][2][*][3] For this reason the inclusion of reducing emissions from land use change is considered essential to achieve the objectives of the UNFCCC.[*][4]

During the negotiations for the Kyoto Protocol, and then in particular its Clean Development Mechanism (CDM), the inclusion of tropical forest management was debated but eventually dropped due to anticipated methodological difficulties in establishing – in particular – additionality and leakage (detrimental effects outside of the project area attributable to project activities). What remained on forestry was "Afforestation and Reforestation", sectoral scope 14 of the CDM. Under this sectoral scope areas of land that had no forest cover since 1990 could be replanted with commercial or indigenous tree species. In its first eight years of operation, a total of 52 projects has been registered under the "Afforestation and Reforestation" scope of the CDM.[*][5] The cumbersome administrative procedures and corresponding high transaction costs are often blamed for this slow uptake.

In response to what many perceived to be a failure to address a major source of global greenhouse gas emissions, the Coalition for Rainforest Nations (CfRN) was estab-lished and in 2005 they proposed to the Conference of the Parties to the UNFCCC a mechanism for considering the reduction of emissions of greenhouse gases stemming from tropical deforestation and forest degradation as a climate change mitigation measure.

30.1 History

30.1.1 REDD

REDD was first discussed in 2005 by the UNFCCC at its 11th session of the Conference of the Parties to the Convention (COP) at the request of Costa Rica and Papua New Guinea, on behalf of the Coalition for Rainforest Nations, when they submitted the document "Reducing Emissions from Deforestation in Developing Countries: Approaches to Stimulate Action",[*][6] with a request to create an agenda item to discuss consideration of reducing emissions from deforestation and forest degradation in natural forests as a mitigation measure. COP 11 entered the request to consider the document as agenda item 6: *Reducing emissions from deforestation in developing countries: approaches to stimulate action.*[*][7]

30.1.2 REDD+

Bali Action Plan

REDD received substantial attention from the UNFCCC – and the attending community – at COP 13, December 2007, where the first substantial decision on REDD+ was adopted, Decision 2/CP.13: "Reducing emissions from deforestation in developing countries: approaches to stimulate action",[*][8] calling for demonstration activities to be reported upon two years later and assessment of drivers of deforestation. Perhaps more interestingly, REDD+ was also referenced in decision 1/CP.13, the "Bali Action Plan", with reference to all five eligible activities for REDD+ (with sustainable management of forests, conservation of forest carbon stocks and enhancement of forest carbon stocks constituting the "+" in REDD+).*[8]

The call for demonstration activities in decision 2/CP.13 led to a very large number of programmes and projects, including the Forest Carbon Partnership Facility (FCPF) of the World Bank, the UN-REDD Programme, and a flurry of smaller projects financed by the Norwegian International Climate and Forest Initiative (NICFI), among many others. All of these were based on interpretation of the very scarce substantive guidance from the UNFCCC. Consequently, many of the projects were only marginally coincident with emerging guidance from the UNFCCC at later sessions.

Definition of main elements

In 2009 at COP 15, decision 4/CP.15: "Methodological guidance for activities relating to reducing emissions from deforestation and forest degradation and the role of conservation, sustainable management of forests and enhancement of forest carbon stocks in developing countries" *[9] provided more substantive information on requirements for REDD+. Specifically, the national forest monitoring system was introduced, with elements of measurement, reporting and verification (MRV). Furthermore, countries were encouraged to develop national strategies, develop domestic capacity, establish reference levels, and establish a participatory approach with "full and effective engagement of indigenous peoples and local communities in (···) monitoring and reporting".

A year later at COP 16 decision 1/CP.16 was adopted.*[10] In section C: "Policy approaches and positive incentives on issues relating to reducing emissions from deforestation and forest degradation in developing countries; and the role of conservation, sustainable management of forests and enhancement of forest carbon stocks in developing countries" the so-called safeguards were introduced, with a reiteration of requirements for the national forest monitoring system. These safeguards were introduced to ensure that implementation of REDD+ at the national level would not lead to detrimental effects for the environment or the local population. It should be noted however that countries are not asked to report on the safeguards themselves, only on how the safeguards are "promoted and supported"; so the mere presence of a regulatory framework should be sufficient to comply with the decision, regardless of how effective it is in actually respecting the safeguards.

In 2011 decision 12/CP.17 was adopted at COP 17: "Guidance on systems for providing information on how safe-

guards are addressed and respected and modalities relating to forest reference emission levels and forest reference levels as referred to in decision 1/CP.16".*[11] Details are provided on preparation and submission of reference levels and guidance on providing information on safeguards.

Warsaw Framework on REDD-plus

In December 2013, COP 19 produced no fewer than seven decisions on REDD+, which are jointly known as the "Warsaw Framework on REDD-plus".*[12] These decisions address a work programme on results-based finance; coordination of support for implementation; modalities for national forest monitoring systems; presenting information on safeguards; technical assessment of reference (emission) levels; modalities for measuring, reporting and verifying (MRV); and information on addressing the drivers of deforestation and forest degradation. Requirements to access to "results-based finance" have been specified: through submission of reports for which the contents have been specified; technical assessment through International Consultation and Analysis (ICA) for which procedures have been specified; and application for results-based finance by developing country Parties to the Green Climate Fund. With these decisions the overall framework for REDD+ implementation appears to be complete, although many details still need to be provided.

COP 20 in December 2014 did not produce any new decisions on REDD+. A reference was made to REDD+ in decision 8/CP.20 "Report of the Green Climate Fund to the Conference of the Parties and guidance to the Green Climate Fund", where in paragraph 18 the COP "*requests* the Board of the Green Climate Fund (...) (b) to consider decisions relevant to REDD-plus", referring back to earlier COP decisions on REDD+.*[13]

30.2 Terminology

The mechanism under discussion by the COP of the UN-FCCC is commonly referred to as "reducing emissions from deforestation and forest degradation", abbreviated to REDD or REDD+. This title and the acronyms, however, are not used by the COP itself.

The original submission by Papua New Guinea and Costa Rica, on behalf of the Coalition for Rainforest Nations, dated 28 July 2005, was entitled "Reducing Emissions from Deforestation in Developing Countries: Approaches to Stimulate Action", exactly as is written here.*[6] COP 11 entered the request to consider the document as agenda item 6: "Reducing emissions from deforestation in developing countries: approaches to stimulate action", again written here exactly as in the official text.*[7] The name for the agenda item was also used at COP 13 in Bali, December 2007. By COP 15 in Copenhagen, December 2009, the scope of the agenda item was broadened to "Methodological guidance for activities relating to reducing emissions from deforestation and forest degradation and the role of conservation, sustainable management of forests and enhancement of forest carbon stocks in developing countries",*[9] moving to "Policy approaches and positive incentives on issues relating to reducing emissions from deforestation and forest degradation in developing countries; and the role of conservation, sustainable management of forests and enhancement of forest carbon stocks in developing countries" by COP 16.*[10] At COP 17 the title of the decision simply referred back to an earlier decision: "Guidance on systems for providing information on how safeguards are addressed and respected and modalities relating to forest reference emission levels and forest reference levels as referred to in decision 1/CP.16".*[11] At COP 19 the titles of decisions 9 and 12 refer back to decision 1/CP.16, paragraph 70 and appendix I respectively, while the other decisions only mention the topic under consideration.*[12]

None of these decisions use an acronym for the title of the agenda item or otherwise; the ubiquitous acronym is thus not coined by the COP of the UNFCCC. Surprisingly therefore, the set of decisions on REDD+ that were adopted at COP 19 in Warsaw, December 2013, were jointly christened the **Warsaw Framework on REDD-plus** in a footnote to the title of each of the decisions.*[12]

All things considered, there should be no confusion on the formal name(s):

- **REDD** originally referred to "reducing emissions from deforestation in developing countries"; the title of the original document on REDD*[7]

- **REDD+** (or **REDD-plus**) refers to "reducing emissions from deforestation *and forest degradation* in developing countries, and the role of *conservation, sustainable management of forests, and enhancement of forest carbon stocks* in developing countries" (emphasis added); the most recent, elaborated terminology used by the COP*[10]

However, the commonly used name outside of the UNFCCC seems to have stuck, perhaps not surprisingly seeing that the original title of the mechanism does not encompass the full scope of forest management options, while the second is quite unwieldy.

30.3 Main elements of REDD+

As a mechanism under the multi-lateral climate change agreement, REDD+ is essentially a vehicle to financially reward developing countries for their verified efforts to reduce emissions and enhance removals of greenhouse gases through a variety of forest management options. As with other mechanisms under the UNFCCC, there are few prescriptions that specifically mandate how to implement the mechanism at national level; the principles of national sovereignty and subsidiarity imply that the UNFCCC can only establish what results it would reward and require that reports are submitted in a certain format and open for review by the Convention. There are certain aspects that go beyond this basic philosophy – such as the so-called safeguards, explained in more detail below – but in essence REDD+ is no more than a set of guidelines on how to report on forest resources and forest management strategies and their results in terms of reducing emissions and enhancing removals of greenhouse gases. However, a set of requirements has been elaborated to ensure that reports from Parties are consistent and comparable and that their content are open to review and in function of the objectives of the Convention.

30.3.1 Policies and measures

In the text of the Convention repeated reference is made to national "policies and measures", the set of legal, regulatory and administrative instruments that Parties develop and implement to achieve the objective of the Convention. These policies can be specific to climate change mitigation or adaptation, or of a more generic nature but with an impact on greenhouse gas emissions. Many of the signatory parties to the UNFCCC have by now established climate change strategies and response measures.

The REDD+ mechanism has a similar, more focused set of policies and measures. Forest sector laws and procedures are typically in place in most countries. In addition, countries have to develop specific national strategies and/or action plans for REDD+.

Of specific interest to REDD+ are the drivers of deforestation and forest degradation. The UNFCCC decisions call on countries to make an assessment of these drivers and to base the policies and measures on this assessment, such that the policies and measures can be directed to where the impact is greatest. Some of the drivers will be generic – in the sense that they are prevalent in many countries, such as increasing population pressure – while others will be very specific to countries or regions within countries.

Countries are encouraged to identify "national circumstances" that impact the drivers: specific conditions within the country that impact the forest resources. Hints for typical national circumstances can be found in preambles to various COP decisions, such as *"Reaffirming that economic and social development and poverty eradication are global priorities"* in the Bali Action Plan,[8] enabling developing countries to prioritize policies like poverty eradication through agricultural expansion or hydropower development over forest protection.

30.3.2 Eligible activities

The decisions on REDD+ enumerate five "eligible activities" that developing countries may implement to reduce emissions and enhance removals of greenhouse gases:

"(a) Reducing emissions from deforestation.

(b) Reducing emissions from forest degradation.

(c) Conservation of forest carbon stocks.

(d) Sustainable management of forests.

(e) Enhancement of forest carbon stocks" .[10]

The first two activities reduce emissions of greenhouse gases and they are the two activities listed in the original submission on REDD+ in 2005 by the Coalition for Rainforest Nations.[6] The three remaining activities constitute the "+" in REDD+. The last one enhances removals of greenhouse gases, while the effect of the other two on emissions or removals is indeterminate but expected to be minimal.

The UNFCCC provides no guidance on what specific actions constitute the eligible activities. Possibly an approach will be adopted as under the CDM: project proponents – in this case Parties to the Convention – can submit documentation on an approach which will be reviewed by a technical committee of the UNFCCC. Upon approval this "approved methodology" will be publicly available to all for its application.

30.3.3 Reference levels

Reference levels are a key component for any national REDD+ program and critical in at least two aspects. Firstly, they will be scrutinized by the international community to assess the quality of the national REDD+ program, in particular with respect to the "fidelity" of the reported emission reductions or enhanced removals. In that sense it establishes the confidence of the international community in the national REDD+ program. Secondly, the reference levels will be the reference against which the achievements of the national REDD+ program will be compared to arrive at

the amount of results-based benefits that countries can expect to receive for their efforts. Setting the reference levels too lax will erode the confidence in the national REDD+ program, while setting them too strict will erode the potential to earn the benefits with which to operate the national REDD+ program. Very careful consideration of all relevant information is therefore of crucial importance.

The requirements and characteristics of reference levels are under the purview of the UNFCCC. Given the wide variety in ecological conditions and country-specific circumstances, these requirements are rather global and every country will have a range of options in its definition of reference levels within its territory.

A reference level (RL) is expressed as an amount, derived by differencing a sequence of amounts over a period of time. For REDD+ purposes the amount is expressed in CO_2-equivalents (CO2e) (see article on global warming potential) of emissions or removals. If the amounts are emissions, the reference level becomes a reference emission level (REL). Reference levels are based on a scope – what is included? – a scale – the geographical area from which it is derived or to which it is applied – and a period over which the reference level is calculated. The scope, the scale and the period can be modified in reference to national circumstances: specific conditions in the country that would call for an adjustment of the basis from which the reference levels are constructed. A reference level can be based on observations or measurements of amounts in the past, in which case it is retrospective, or it can be an expectation or projection of amounts into the future, in which case it is prospective.[14]

Reference levels have to have national coverage, but they may be composed from a number of sub-national reference levels. As an example, forest degradation may have a reference emission level for commercial selective logging and one for extraction of minor timber and firewood for subsistence use by rural communities. Effectively, every identified driver of deforestation or forest degradation has to be represented in one or more reference emission level(s). Similarly for reference levels for enhancement of carbon stocks, there may be a reference level for plantation timber species and one for natural regeneration, possibly stratified by ecological region or forest type.

Details on the reporting and technical assessment of reference levels is given in Decision 13/CP.19.[12]

30.3.4 Monitoring: measurement, reporting and verification

In Decision 2/CP.15 of the UNFCCC countries are requested to develop national forest monitoring systems

(NFMS) that support the functions of measurement, reporting and verification (MRV) of actions and achievements of the implementation of REDD+ activities.*[9] NFMS is the key component in the management of information for national REDD+ programs. A fully functional monitoring system can go beyond the requirements posted by the UNFCCC to include issues such as a registry of projects and participants, and evaluation of program achievements and policy effectiveness. It may be purpose-built, but it may also be integrated into existing forest monitoring tools.

Measurements are suggested to be made using a combination of remote sensing and ground-based observations. Remote sensing is particularly suited to the assessment of areas of forest and stratification of different forest types. Ground-based observations involve forest surveys to measure the carbon pools used by the IPCC, as well as other parameters of interest such as those related to safeguards and eligible activity implementation.*[15]

The reporting has to follow the guidance of the IPCC, in particular the "Good Practice Guidance for Land Use, Land-Use Change and Forestry",*[16] which includes reporting templates to be included in National Communications of Parties to the UNFCCC. Included in the guidance are standard measurements protocols and analysis procedures which greatly impact the measurement systems that countries need to establish. The actual reporting of REDD+ results is not going through the National Communications, however, but through the Biennial Update Reports (BURs).*[11]

Verification is an independent, external process that is managed by the Secretariat to the UNFCCC; countries need to facilitate the requirements of verification. Verification goes through a process of International Consultation and Analysis (ICA), which is effectively a peer-review by a team composed of an expert from an Annex I Party and an expert from a non-Annex I Party which "will be conducted in a manner that is nonintrusive, non-punitive and respectful of national sovereignty".*[11] This "technical team of experts shall analyse the extent to which:

(a) There is consistency in methodologies, definitions, comprehensiveness and the information provided between the assessed reference level and the results of the implementation of the [REDD+] activities (...);

(b) The data and information provided in the technical annex is transparent, consistent, complete and accurate;

(c) The data and information provided in the technical annex is consistent with the [UNFCCC] guidelines (...);

(d) The results are accurate, to the extent possible." *[12]

30.3.5 Safeguards

In response to concerns over the potential for misuse and misappropriation of REDD+ the UNFCCC established a list of safeguards that countries need to "promote and support" in order to guarantee the correct and lasting generation of results from the REDD+ mechanism. These safeguards are:

"(a) That actions complement or are consistent with the objectives of national forest programmes and relevant international conventions and agreements;

(b) Transparent and effective national forest governance structures, taking into account national legislation and sovereignty;

(c) Respect for the knowledge and rights of indigenous peoples and members of local communities, by taking into account relevant international obligations, national circumstances and laws, and noting that the United Nations General Assembly has adopted the United Nations Declaration on the Rights of Indigenous Peoples;

(d) The full and effective participation of relevant stakeholders, in particular indigenous peoples and local communities;

(e) That actions are consistent with the conservation of natural forests and biological diversity, ensuring that the actions are not used for the conversion of natural forests, but are instead used to incentivize the protection and conservation of natural forests and their ecosystem services, and to enhance other social and environmental benefits;

(f) Actions to address the risks of reversals;

(g) Actions to reduce displacement of emissions" .*[10]

Countries do not have to report on the safeguards themselves – e.g. how many local communities are effectively participating – but only demonstrate how the safeguards are respected. This could come in the form, for instance, of explaining the legal and regulatory environment with regards to the recognition, inclusion and engagement of Indigenous Peoples.

Decision 12/CP.19 established that the "summary of information" on the safeguards will be provided in the National Communications to the UNFCCC, which for developing country Parties will be once every four years. Additionally, and on a voluntary basis, the summary of information may be posted on the UNFCCC REDD+ web platform.*[12]*[17]

30.3.6   Not REDD+

The REDD+ mechanism is currently still under discussion by the UNFCCC. All pertinent issues that comprise REDD+ are exclusively those that are included in the decisions of the COP, as indicated in the above sections. There is, however, a large variety of concepts and approaches that are labelled (as being part of) REDD+ by their proponents, either being a substitute for UNFCCC decisions or complementary to those decisions. Below follows a – no doubt, incomplete – list of such concepts and approaches.

- **Project-based REDD+, voluntary market REDD+**. Many organizations promote REDD+ projects at the scale of a forest area (e.g. large concession, National Park), analogous to AR-CDM projects under the Kyoto Protocol, with reduction of emissions or enhancement of removals vetted by an external organization using a standard established by some party (e.g. CCBA, VCS) and with carbon credits traded on the international voluntary carbon market. However, REDD+ is by definition national (Decisions 4/CP.15 and 1/CP.16 consistently refer to national strategies and action plans and national monitoring, with sub-national coverage allowed as an interim measure only*[9]*[10]) and no agreement has yet been reached by the UNFCCC on how to convert verified net emission reductions or enhanced removals of greenhouse gases into a financial instrument.

- **Benefit distribution**. The UNFCCC decisions on REDD+ are silent on the issue of rewarding countries and participants for their verified net emission reductions or enhanced removals of greenhouse gases. It is not very likely that specific requirements for sub-national implementation of the distribution of benefits will be adopted, as this will be perceived to be an issue of national sovereignty. Generic guidance may be provided, using language similar to that of the safeguards, such as "result-based finance has to accrue to local stakeholders" without being specific on percentages retention for management, identification of stakeholders, type of benefit or means of distribution. Countries may decide to channel any benefits through an existing program on rural development, for instance, provide additional services (e.g. extension, better market access, training, seedlings) or pay local stakeholders directly.

- **FPIC**. Free, prior and informed consent is a required procedure under UN-REDD, with the global Programme having mandated its use in countries benefiting from its support.*[18] The REDD+ mechanism has no such requirement, however, although the safeguard on respect for the knowledge and rights of indigenous peoples and members of local communities *notes* "that the United Nations General Assembly has adopted the United Nations Declaration on the Rights of Indigenous Peoples" (UNDRIP).*[10] Article 19 of UNDRIP requires that "States shall consult and cooperate in good faith with the indigenous peoples concerned through their own representative institutions in order to obtain their free, prior and informed consent before adopting and implementing legislative or administrative measures that may affect them". This article is interpreted by the UN-REDD Programme in their "Guidelines on Free, Prior and Informed Consent" to mean that every, or at least many, communities need to provide their consent before any REDD+ activities can take place.*[18] It may be obvious that most of the elements of a standard FPIC process are required to make a local participatory forest management activity, such as any of the eligible REDD+ activities, successful and sustainable, but as such FPIC is not required to implement REDD+.

- **Leakage**. Leakage is a term that is often used in project-based REDD+. The term originates from Afforestation/Reforestation projects under the CDM of the Kyoto Protocol where it is assessed to quantify effects of the project outside of the project area. Under REDD+ the term is not used and makes no sense either. Since REDD+ has national scope there can be no leakage domestically, once full national coverage is achieved. Internationally there can be no leakage either because all countries are required to report on forest resources, through REDD+ or otherwise, so an increase in logging in one country to offset reduced logging in another country is registered in the country where the timber was sourced from. REDD+ does use the term "displacement of emissions" but this is to be understood as displacement of emissions between sectors, such as replacing wood fires with kerosene stoves (AFOLU to energy) or construction with wood for construction with concrete, cement and bricks (AFOLU to industry).

30.4 REDD+ as a climate change mitigation measure

Deforestation and forest degradation account for 17-29% of global greenhouse gas emissions,[*][1][*][2][*][3] the reduction of which is estimated to be one of the most cost-efficient climate change mitigation strategies.[*][19][*][20] Regeneration of forest on degraded or deforested lands can remove CO_2 from the atmosphere through the build-up of biomass, making forest lands a sink of greenhouse gases. The REDD+ mechanism addresses both issues of emission reduction and enhanced removal of greenhouse gases.

30.4.1 Reducing emissions

Emissions of greenhouse gases from forest land can be reduced by slowing down the rates of deforestation and forest degradation, obviously covered by the first two of the REDD+ *eligible activities*. Another option would be some form of reduced impact logging in commercial logging, under the REDD+ *eligible activity* of sustainable management of forests.

30.4.2 Enhancing removals

Removals of greenhouse gases (specifically CO_2) from the atmosphere can be achieved through various forest management options, such as replanting degraded or deforested areas or enrichment planting, but also by letting forest land regenerate naturally. Care must be taken to differentiate between what is a purely ecological process of regrowth and what is induced or enhanced through some management intervention.

30.4.3 REDD+ and the carbon market

In 2009, at COP-15 in Copenhagen, the Copenhagen Accord was reached, noting in section 6 the recognition of the crucial role of REDD and REDD+ and the need to provide positive incentives for such actions by enabling the mobilization of financial resources from developed countries. The Accord goes on to note in section 8 that the collective commitment by developed countries for new and additional resources, including forestry and investments through international institutions, will approach USD 30 billion for the period 2010 - 2012.[*][21]

The Green Climate Fund (GCF) was established at COP-17 to function as the financial mechanism for the UNFCCC, so including for REDD+ finance. The Warsaw Framework on REDD-plus makes various references to the GCF, instructing developing country Parties to apply to the GCF for *result-based finance*.[*][12]

30.5 Implementing REDD+

Decision 1/CP.16, paragraph 73, suggests that national capacity for implementing REDD+ is built up in phases, "beginning with the development of national strategies or action plans, policies and measures, and capacity-building, followed by the implementation of national policies and measures and national strategies or action plans that could involve further capacity-building, technology development and transfer and results-based demonstration activities, and evolving into results-based actions that should be fully measured, reported and verified".[*][10] The initial phase of the development of national strategies and action plans and capacity building is typically referred to as the "Readiness phase" (a term like *Reddiness* is also encountered).

There is a very substantial number of REDD+ projects globally and this section lists only a selection. One of the more comprehensive online tools with up-to-date information on REDD+ projects is the Voluntary REDD+ Database.

30.5.1 Readiness activities

Most REDD+ activities or projects implemented since the call for demonstration activities in Decision 2/CP.13 December 2007[*][8] are focused on readiness, which is not surprising given that REDD+ and its requirements were completely new to all developing countries.

- **UN-REDD Programme** UNDP, UNEP and FAO jointly established the UN-REDD Programme in 2007, a partnership aimed at assisting developing countries in addressing certain measures needed in order to effectively participate in the REDD+ mechanism. These measures include capacity development, governance, engagement of Indigenous Peoples and technical needs. The initial set of supported countries were Bolivia, Democratic Republic of Congo, Indonesia, Panama, Papua New Guinea, Paraguay, Tanzania, Vietnam, and Zambia. By March 2014 the Programme counted 49 participants, 18 of which are receiving financial support to kick start or complement a variety of national REDD+ readiness activities.[*][22] The other 31 *partner countries* may receive *targeted support* and *knowledge sharing*, be invited to attend meetings and training workshops, have *observer status* at the Policy Board meetings, and "may be invited to submit a request to receive funding for a National Programme in the future, if selected through a set of criteria to prioritize funding for new countries approved by the Policy Board".*[23] The Programme operates in six work areas:*[24]

1. MRV and Monitoring (led by FAO)

2. National REDD+ Governance (UNDP)

3. Engagement of Indigenous Peoples, Local Communities and Other Relevant Stakeholders (UNDP)

4. Ensuring multiple benefits of forests and REDD+ (UNEP)

5. Transparent, Equitable and Accountable Management of REDD+ Payments (UNDP)

6. REDD+ as a Catalyst for Transformations to a Green Economy (UNEP)

- **Forest Carbon Partnership Facility** The World Bank plays an important role in the development of REDD+ activities since its inception. The Forest Carbon Partnership Facility (FCPF) was presented to the international community at COP-11 in Bali, December 2007. Recipient countries can apply $3.6 million towards: the development of national strategies; stakeholder consultation; capacity building; development of reference levels; development of a national forest monitoring system; and social and environmental safeguards analysis.*[25] Those countries that successfully achieve a state of readiness can apply to the related Carbon Fund, for support towards national implementation of REDD+.*[26]

- **Norwegian International Climate and Forest Initiative** At the 2007 Bali Conference, the Norwegian government announced their International Climate and Forests Initiative (NICFI), which provided US$1 billion towards the Brazilian REDD scheme*[27] and US$500 million towards the creation and implementation of national-based, REDD+ activities in Tanzania.*[28] In addition, with the United Kingdom, $200 million was contributed towards the Congo Basin Forest Fund to aid forest conservation activities in Central Africa.*[29] In 2010, Norway signed a Letter of Intent with Indonesia to provide the latter country with up to US$1 billion "assuming that Indonesia achieves good results".*[30]

- **ITTO** The International Tropical Timber Organization (ITTO) has launched a thematic program on REDD+ and environmental services with an initial funding of US$3.5 million from Norway. In addition, the 45th session of the ITTO Council held in November 2009, recommended that efforts relating REDD+

should focus on promoting "sustainable forest management".

- **Finland** In 2009, the Government of Finland and the Food and Agriculture Organization of the United Nations signed a US$17 million partnership agreement to provide tools and methods for multi-purpose forest inventories, REDD+ monitoring and climate change adaptation in five pilot countries: Ecuador, Peru, Tanzania, Viet Nam and Zambia.*[31] As part of this programme, the Government of Tanzania will soon complete the country's first comprehensive forest inventory to assess its forest resources including the size of the carbon stock stored within its forests. A forest soil carbon monitoring program to estimate soil carbon stock, using both survey and modelling-based methods, has also been undertaken.*[32]

- **Australia** Australia established a A$200 million International Forest Carbon Initiative, focused on developing REDD+ activities in its vicinity, i.e., in areas like Indonesia, and Papua New Guinea.*[33]

- **Interim REDD+ Partnership** In 2010, national governments of developing and developed countries joined efforts to create the Interim REDD+ Partnership as means to enhance implementation of early action and foster fast start finance for REDD+ actions.*[34]

30.5.2 Implementation phase

Arguably, some countries are already implementing aspects of a national forest monitoring system and activities aimed at reducing emissions and enhancing removals that go beyond REDD+ readiness. Costa Rica, for instance, has applied to the FCPF Carbon Fund and its application was successful, indication that by FCPF standards Costa Rica has completed REDD+ readiness. Other countries with advanced REDD+ frameworks going beyond REDD+ readiness include Brazil and Mexico.

30.5.3 Results-based actions

By March 2014 no developing countries had entered into the phase of results-based actions that are fully measured, reported and verified.

30.6 Concerns

Since the first discussion on REDD+ in 2005 and particularly at COP-13 in 2007 and COP-15 in 2009, many concerns have been voiced on various aspects of REDD+. The COP has responded by establishing the safeguards for REDD+, although these are widely criticized for being too generic, non-enforceable and summary rather than a specific set of requirements for participation in the REDD+ mechanism.

Prior to full-scale implementation many challenges are still to be solved. How will the REDD+ mechanism link to existing national development strategies? How can forest communities and indigenous peoples participate in the design, implementation, monitoring and evaluation of national REDD+ programmes? How will REDD+ be funded, and how will countries ensure that benefits are distributed equitably among all those who manage the forests? Finally, how will the amounts of reduced emissions and enhanced removals as a result of REDD+ be monitored?

30.6.1 Natural forests vs. high-density plantations

Safeguard (e): That actions are consistent with the conservation of natural forests and biological diversity, ensuring that the [REDD+] actions (...) are not used for the conversion of natural forests, but are instead used to incentivize the protection and conservation of natural forests and their ecosystem services, and to enhance other social and environmental benefits. Footnote to this safeguard: Taking into account the need for sustainable livelihoods of indigenous peoples and local communities and their interdependence on forests in most countries, reflected in the United Nations Declaration on the Rights of Indigenous Peoples, as well as the International Mother Earth Day.

The UNFCCC does not define what constitutes a forest; it only requires that Parties communicate to the UNFCCC how they define a forest, but suggesting to use a definition in terms of minimal area, minimal crown coverage and minimal height at maturity of perennial vegetation.

While there is a safeguard against the conversion of natural forest, developing country Parties are free to include plantations of commercial tree species (including exotics like Eucalyptus spp., Pinus spp., Acacia spp.), agricultural tree crops (e.g. rubber, mango, cocoa, citrus), or even non-tree species such as palms (oil palm, coconut, dates) and bamboo (a grass). Some opponents of REDD+ argue that this lack of a clear distinction is no accident. Defining a forest simply in terms of tree cover - rather than complex ecosystems and the livelihoods of peoples interacting with them – has long been used as a cover for the expansion of industrial-scale plantations. The most plausible explanation, arguably, is that commercial interests take precedence over environmental and social objectives in the shaping of REDD+ policy.

Similarly, there is no consensus on a definition for forest degradation.*[35] The IPCC has come up with a number of suggestions, again leaving countries the option to select that definition which is most convenient.

A national REDD+ strategy need not refer solely to the establishment of national parks or protected areas; by the careful design of rules and guidelines, REDD+ could include land use practices such as shifting cultivation by indigenous communities and reduced-impact-logging, provided sustainable rotation and harvesting cycles can be demonstrated.*[36] Some argue that this is opening the door to logging operations in primary forests, displacement of local populations for "conservation", increase of tree plantations.

Achieving multiple benefits, for example the conservation of biodiversity and ecosystem services (such as drainage basins), and social benefits (for example income and improved forest governance) is currently not addressed, beyond the inclusion in the safeguard.

30.6.2 Land tenure, carbon rights and benefit distribution

According to some critics, REDD+ is another extension of green capitalism, subjecting the forests and its inhabitants to new ways of expropriation and enclosure at the hands of polluting companies and market speculators. So-called "carbon cowboys" - unscrupulous entrepreneurs who attempt to acquire rights to carbon in rainforest - have signed on indigenous communities to unfair contracts, often with a view to on-selling the rights to investors for a quick profit. In 2012 an Australian businessman operating in Peru was revealed to have signed 200-year contracts with an Amazon tribe, the Yagua, many members of which are illiterate, giving him a 50 per cent share in their carbon resources. The contracts allow him to establish and control timber projects and palm oil plantations in Yagua rainforest.*[37]

There are risks that the local inhabitants and the communities that live in the forests will be bypassed and that they won't be consulted and so they won't actually receive any revenues.*[38] Fair distribution of REDD+ benefits will not be achieved without a *prior* reform in forest governance and more secure tenure systems in many countries.*[39] So how can the benefits from REDD+ be distributed to forest communities in a just, equitable way that minimizes capture of the benefits by national governments or local elites?*[40]

The UNFCCC has repeatedly called for *full and effective participation of Indigenous Peoples and local communities* without becoming any more specific. The ability of local communities to effectively contribute to REDD+ field activities and the measurement of forest properties for estimating reduced emissions and enhanced emissions of greenhouse gases has been clearly demonstrated in various countries.[*][41]

In project-based REDD+, some projects are unaccountable and dodgy companies are taking advantage of the low governance.[*][42]

30.6.3 Indigenous Peoples

Safeguard (c): Respect for the knowledge and rights of indigenous peoples and members of local communities, by taking into account relevant international obligations, national circumstances and laws, and noting that the United Nations General Assembly has adopted the United Nations Declaration on the Rights of Indigenous Peoples; Safeguard (d): The full and effective participation of relevant stakeholders, in particular indigenous peoples and local communities, in the [REDD+] actions (...) [and when developing and implementing national strategies or action plans];

Indigenous peoples are important stakeholders in REDD+ as they typically live inside forest areas and/or have their livelihoods (partially) based on exploitation of forest resources. The International Indigenous Peoples Forum on Climate Change (IIPFCC) was explicit at the Bali climate negotiations in 2007:

> *REDD/REDD+ will not benefit Indigenous Peoples, but in fact will result in more violations of Indigenous Peoples' rights. It will increase the violation of our human rights, our rights to our lands, territories and resources, steal our land, cause forced evictions, prevent access and threaten indigenous agricultural practices, destroy biodiversity and cultural diversity and cause social conflicts. Under REDD/REDD+, states and carbon traders will take more control over our forests.[*][43]*

Putting a commercial value on forests neglects the spiritual value they hold for Indigenous Peoples and local communities.[*][2]

Indigenous Peoples protested in 2008 against the United Nations Permanent Forum on Indigenous Issues final report on climate change and a paragraph that endorsed REDD+; this was captured in a video entitled "the 2nd May Revolt"
·

Indigenous Peoples' groups in Panama broke off their collaboration with the national UN-REDD Programme in 2012 over allegations of a failure of the government to properly respect the rights of the indigenous groups.

Some grassroots organizations are working to develop REDD+ activities with communities and developing benefit-sharing mechanisms to ensure REDD+ funds reach rural communities as well as governments. Examples of these include Plan Vivo projects in Mexico, Mozambique and Cameroon.[*][44]

30.6.4 REDD+ in the carbon market

When REDD+ was first discussed by the UNFCCC, no indication was given on how developing countries would be financially compensated for their efforts to implement REDD+ to reduce emissions and enhance removals of greenhouse gases from forests. In the absence of guidance from the COP, two options were debated by the international community at large:

1. a market-based approach;

2. a fund-based approach where Annex I countries would deposit substantial amounts of money into a fund administered by some multi-lateral entity.

Under the market-based approach, REDD+ would act as an "offset scheme" in which verified results-based actions translate into some form of carbon credits, more-or-less analogous to the market for Certified Emission Reductions (CER) under the CDM of the Kyoto Protocol. Such carbon credits could then offset emissions in the country or company of the buyer of the carbon credits. This would require Annex I countries to agree to deeper cuts in emissions of greenhouse gases in order to create a market for the carbon credits from REDD+, which is unlikely to happen soon given the current state of negotiations in the COP, but even then there is the fear that the market will be flooded with carbon credits, depressing the price to levels where REDD+ is no longer an economically viable option.[*][45][*][46] Some developing countries, such as Brazil and China, maintain that developed countries must commit to real emissions reductions, independent of any offset mechanism.[*][47]

Recent studies indicate that an offset approach based on projects would significantly increase the transaction costs associated to REDD+ and would actually be the weakest alternative for a national REDD+ architecture as regards effectiveness, efficiency, its capacity to deliver co benefits (like development, biodiversity or human rights) and its overall political legitimacy.[*][48]

Since COP-17, however, it has become clear that the COP is opting for a fund-based financing of REDD+, with the newly established Green Climate Fund as the trustee for management of the Fund and disbursement of result-based finance to developing countries that submit verified reports of emission reductions and enhanced removals of greenhouse gases.*[11]*[12] This fund is only available to developing country Parties to the UNFCCC, however, so any REDD+ projects in the voluntary carbon market would still require other means to market verified emission reductions.

30.6.5 Institutional technocrats vs. stakeholders

While the COP decisions emphasize national ownership and stakeholder consultation, there are concerns that some of the larger institutional organizations are driving the process, in particular outside of the *one Party, one vote* realm of multi-lateral negotiations under the UNFCCC. For example, the World Bank and the UN-REDD Programme, the two largest sources of funding and technical assistance for readiness activities and therefore unavoidable for most developing countries, place requirements upon recipient countries that are arguably not mandated or required by the COP decisions.

Although the World Bank declares its commitment to fight against climate change, many civil society organisations and grassroots movements around the world view with scepticism the processes being developed under the various carbon funds. Among some of the most worrying reasons are the weak (or inexistent) consultation processes with local communities; the lack of criteria to determine when a country is ready to implement REDD+ projects (readiness); the negative impacts such as deforestation and loss of biodiversity (due to fast agreements and lack of planning); the lack of safeguards to protect Indigenous Peoples' rights; and the lack of regional policies to stop deforestation. A growing coalition of civil society organization, social movement, and other actors critical of REDD+ emerged between 2008 and 2011, criticizing the mechanism on climate justice grounds.*[49] During the UN climate negotiations in Copenhagen (2009) and Cancun (2010) strong civil society and social movements coalitions formed a strong front to fight the World Bank out of the climate.

ITTO has been criticized for appearing to support above all the inclusion of forest extraction inside REDD+ under the guise of "sustainable management" in order to benefit from carbon markets while maintaining business-as-usual.*[43]

30.7 See also

- Deforestation
- Deforestation by region
- Emissions trading
- Environmental crime
- Illegal logging
- International Work Group for Indigenous Affairs
- Kyoto Protocol
- CDM excluding Forest Conservation
- Participatory monitoring
- Political corruption
- Tree credits
- United Nations Forum on Forests

30.8 References

30.8.1 Notes

[1] Philip Fearnside (2000). "Global warming and tropical land-use change: Greenhouse gas emissions from biomass burning, decomposition and soils in forest conversion, shifting cultivation and secondary vegetation". *Climatic Change* **46**: 115–158. doi:10.1023/a:1005569915357.

[2] Myers,Erin C. (Dec 2007). "Policies to Reduce Emissions from Deforestation and Degradation (REDD) in Tropical Forests" (PDF). *Resources Magazine*: 7. Retrieved 2009-11-24.

[3] G.R. van der Werf, D. C. Morton, R. S. DeFries, J. G. J. Olivier, P. S. Kasibhatla, R. B. Jackson, G. J. Collatz and J. T. Randerson (Nov 2009). "CO2 emissions from forest loss". *Nature Geoscience* **2** (11): 737–738. doi:10.1038/ngeo671.

[4] Butler, Rhett (Aug 2009). "Big REDD". *Washington Monthly* **41**: 2.

[5] UNFCCC CDM project search page, accessed 2014-02-28

[6] "Microsoft Word - cpsc1.doc" (PDF). Retrieved 2014-02-21.

[7] "UNFCCC document FCCC/CP/2005/5" (PDF). Retrieved 2014-02-21.

[8] "Microsoft Word - cp6a1 reissued.doc" (PDF). Retrieved 2014-02-21.

[9] "Microsoft Word - cp11add.1.doc" (PDF). Retrieved 2014-02-21.

[10] "Microsoft Word - cp7a1.doc" (PDF). Retrieved 2014-02-21.

[11] http://unfccc.int/resource/docs/2011/cop17/eng/09a02.pdf

[12] "UNFCCC document FCCC/CP/2013/10/Add.1" (PDF). Retrieved 2014-02-21.

[13] "Report of the Conference of the Parties on its twentieth session, held in Lima from 1 to 14 December 2014" (PDF). 2015-02-02. Retrieved 2015-04-20.

[14] Guidelines for REDD+ Reference Levels: Principles and Recommendations

[15] Integrating remote-sensing and ground-based observations for estimation of emissions and removals of greenhouse gases in forests

[16] Good Practice Guidance for Land Use, Land-Use Change and Forestry

[17] UNFCCC REDD+ web platform

[18] UN-REDD Guidelines on Free, Prior and Informed Consent

[19] Stern Review

[20] Eliasch Review

[21] "Copenhagen Accord of 18 December 2009" (PDF). UN-FCC. 2009. Retrieved 2009-12-28.

[22] "UN-REDD Programme - Support to Partner Countries". Un-redd.org. Retrieved 2014-03-02.

[23]

[24] UN-REDD Programme Strategy 2011-2015

[25] FCPF template to apply for funding.

[26] FCPF Carbon Fund As the first country to apply for funding under the FCPF Carbon Fund, Costa Rica submitted its revised *ER-PIN* on 15 February 2013. Apparently only five submissions are currently envisaged.

[27] Amazon Fund website

[28] "The Government of Norway's International Climate and Forest Initiative". Ministry of Environment. May 19, 2009. Retrieved 2009-11-23.

[29] Butler, Rhett (July 22, 2009). "Are We on The Brink of Saving Rainforests?". Retrieved 2009-11-23.

[30]

[31] "Support to forest monitoring and assessment". Fao.org. Retrieved 2013-12-18.

[32] "Soil carbon monitoring using surveys and modelling". Fao.org. Retrieved 2013-12-18.

[33] "Reducing Emission from Deforestation and Forest Degradation in Developing Countries" (PDF). UNFCC. Retrieved 2009-11-23.

[34] "REDD+ Partnership website". Reddpluspartnership.org. Retrieved 2013-12-18.

[35] "CDM Carbon Sink Tree Plantations: Insights into Sustainability Issues". Thinktosustain.com. 2011-05-20. Retrieved 2013-12-18.

[36] Mathai, J. (5 October 2009). "Seeing REDD over deforestation". http://www.peat-portal.net/newsmaster.cfm?&menuid=38&action=view&retrieveid=1060

[37] Stephen Rice and Liam Bartlett (6 July 2012). "The Carbon Cowboy". 60 Minutes, Nine Network Australia. Retrieved 2012-11-25.

[38] Espinoza Llanos, Roberto and Feather, Conrad (Nov 2011). "The reality of REDD+ in Peru: Between theory and practice - Indigenous Amazonian Peoples' analyses and alternatives" (PDF). AIDESEP and Forest Peoples Programme. Retrieved 2009-11-23.

[39] Vickers, Ben (Apr 2008). "REDD: a Steep learning Curve" (PDF). Asia-Pacific Forestry Week. Retrieved 2009-11-23.

[40] Peskett, Leo; Huberman,David; Bowen-Jones,Evan; Edwards,Guy; Brown,Jessica (Sep 2008). "Making REDD Work for the Poor" (PDF). Poverty Environment Partnership. Retrieved 2009-11-23.

[41] In a multi-year research project the engagement of local communities in REDD+ field activity implementation in 7 developing countries was studied and found to be highly effective. See multiple papers at under the Resources & Publications link on the left.

[42] "Carbon offsetting scheme open to corruption, report warns". Abc.net.au. 2010-11-01. Retrieved 2013-12-18.

[43] "No Redd!" (PDF). Noredd.makenoise.org. November 2011. Retrieved 2014-12-11.

[44] "Plan Vivo " Improving livelihoods, conserving and restoring ecosystems". Planvivo.org. Retrieved 2013-12-18.

[45] Butler, Rhett (Mar 5, 2008). "Why Europe Torpedoed the REDD Forest-For-Carbon Credits Initiative". Retrieved 2009-11-23.

[46] Karsenty, Alain (Nov 2009). "What the (carbon) market cannot do...". CIRAD. Retrieved 2012-02-15.

[47] von der Goltz, Jan (August 10, 2009), *High Stakes in a Complex Game: A Snapshot of the Climate Change Negotiating Positions of Major Developing Country Emitters* (PDF), Center for Global Development

[48] Arild Vatn, Paul Vedeld (April 2011). "Getting Ready! A Study of National Governance Structuresfor REDD+" (PDF). NORAGRIC. Retrieved 2009-11-23.

[49] den Besten, Jan Willem; Arts, Bas; Verkooijen, Patrick (2014). "The evolution of REDD+: an analysis of discursive-institutional dynamics". *Environmental Science & Policy* 35: 40–48. doi:10.1016/j.envsci.2013.03.009.

30.8.2 Further reading

- Lu, H.; Wang, X.; Zhang, Y.; Yan, W.; Zhang, J. (2012). "Modelling Forest Fragmentation and Carbon Emissions for REDD plus". *Procedia Engineering* **37**: 333. doi:10.1016/j.proeng.2012.04.249.

- Probert, C., Sharrock, S. and Ali, N. 2011. A REDD+ manual for botanic gardens Botanic Gardens Conservation International (BGCI)

30.9   External links

- UNFCCC REDD Web Platform

- The REDD Desk - A Collaborative Resource for REDD Readiness

- REDD+ Partnership, including financing database

- Forest Carbon Partnership Facility, hosted by the World Bank

- REDD+ profile on database of Market Governance Mechanisms

- New method for interpreting satellite imagery for REDD, developed by Gregory Asner

- UN-REDD Programme

- The Plan Vivo System: A framework for Community REDD+ schemes

- Code REDD: A campaign to promote REDD+ projects and the corporations who have pledged support

- List of REDD projects in Indonesia

- Norway's International Climate and Forest Initiative

- Land Change Modeling for REDD blog

- REDD-Monitor - Critical analysis and news about REDD

- REDD-plus - News Views and Analysis on REDD-plus

- REDD+ - The Online Library - read everything about Deforestation, Sustainable Forest Management, Illegal Logging and a lot more...

- No REDD, A Reader

- Carbon Trade Watch

- Is REDD the New Green? Indigenous Groups Resist Market-Based Forestry Scheme - video report by *Democracy Now!*

- What's happened to Guyana's rainforest deal with Norway?

- The Carbon Cowboy - report by Liam Bartlett and Stephen Rice *60 Minutes (Australia)*

Chapter 31

Secondary forest

The forest in Stanley Park, Vancouver, Canada is generally considered to have second and third growth characteristics. This photo shows regeneration, a tree growing out of the stump of another tree that was felled in 1962 by the remnants of Typhoon Freda.

A **secondary forest** (or **second-growth forest**) is a forest or woodland area which has re-grown after a major disturbance such as fire, insect infestation, timber harvest or windthrow, until a long enough period has passed so that the effects of the disturbance are no longer evident. It is distinguished from an old-growth forest (primary or primeval forest), which has not undergone such disruptions, as well as third-growth forests that result from severe disruptions in second growth forests.

31.1   Description

Depending on the forest, the development of primary characteristics may take anywhere from a century to several millennia. Hardwood forests of the eastern United States, for example, can develop primary characteristics in one or two generations of trees, or 150-500 years. Often the disruption is the result of human activity, such as logging, but natural phenomena that produce the same effect are often included in the definition. Secondary forests tend to have trees closer spaced than primary forests and contain less undergrowth than primary forests. Secondary forests typically were thought to lack biodiversity compared to primary forests, however this has been challenged in recent years. Usually, secondary forests have only one canopy layer, whereas primary forests have several.

Secondary forestation is common in areas where forests have been lost by the slash-and-burn method, a component of some shifting cultivation systems of agriculture. Secondary forests may also arise from forest that has been harvested heavily or over a long period of time, forest that is naturally regenerating from fire and from abandoned pastures or areas of agriculture. It takes a secondary forest typically forty to 100 years to begin to resemble the original old-growth forest; however, in some cases a secondary forest will not succeed, due to erosion or soil nutrient loss in certain tropical forests.

Secondary forests re-establish by the process of succession. Openings created in the forest canopy allow sunlight to reach the forest floor. An area that has been cleared will first be colonized by pioneer species. Even though some species loss may occur with primary forest removal, a secondary forest can protect the watershed from further erosion and provides habitat. Secondary forests may also buffer edge effects around mature forest fragments and increase connectivity between them. They may also be a source of wood and other forest products.

Today most of the forest of the United States, the eastern part of North America and Europe consist of secondary forest.

31.2 Rainforests

In the case of tropical rainforests, where soil nutrient levels are characteristically low, the soil quality may be significantly diminished following the removal of primary forest. In Panama, growth of new forests from abandoned farmland exceeded loss of primary rainforest in 1990.[*][1] However, due to the diminished quality of soil, among other factors, the presence of a significant majority of primary forest species fail to recover in these second-growth forests.

31.3 Notes

[1] "New Jungles Prompt a Debate on Rain Forests" article by Elisabeth Rosenthal in *The New York Times* January 29, 2009

31.4 References

- CIFOR Secondary Forest
- FAO Forestry
- World Resource Institute

31.5 External links

- M. van Breugel, 2007, Dynamics of secondary forests. PhD Thesis Wageningen University. ISBN 978-90-8504-693-6

- Uzay. U Sezen, 2007, Parentage analysis of a regenerating palm tree in a tropical second-growth forest. Ecological Society of America, Ecology 88: 3065-3075.

Chapter 32

Seed bombing

Seed bombs used as a protest tool at the March Against Monsanto, 2013

Seed bombing or **aerial reforestation**[*][1] is a technique of introducing vegetation to land by throwing or dropping compressed bundles of soil containing live vegetation (seed balls). Often, seed bombing projects are done with arid or off-limits (for example, privately owned) land.

32.1 History

The term "seed grenade" was first used by Liz Christy in 1973 when she started the "Green Guerrillas". The first seed grenades were made from balloons filled with tomato seeds, and fertilizer.[*][2] They were tossed over fences onto empty lots in New York City in order to make the neighborhoods look better. It was the start of the guerrilla gardening movement.

The earliest records of aerial reforestation date back from 1930. In this period, planes were used to distribute seeds over certain inaccessible mountains in Honolulu after forest fires.[*][1]

Seed bombing is also widely used in Africa; where they are put in barren or simply grassy areas. With technology expanding, the contents of a seed bomb are now placed in a biodegradable container and "bombed" grenade-style onto the land. As the sprout grows, the container biodegrades into the soil. The process is usually done as a large-scale project with hundreds dropped in a single area at any one time. Provided enough water, adequate sunlight, and low competition from existing flora and fauna, seed-bombed barren land could be host to new plants in as little as a month.

In 1987, Lynn Garrison created the Haitian Aerial Reforestation Project (HARP) in which tons of seed would be scattered from specially modified aircraft. The seeds would be encapsulated in an absorbent material. This coating would contain fertilizer, insecticide/animal repellent and, perhaps a few vegetable seeds. Haiti has a bimodal rainy season, with precipitation in spring and fall. The seeds are moistened a few days before the drop, to start germination. Tons of seed can be scattered across areas in the mountains, inaccessible to hand-planting projects.

Another project idea was to use C-130 aircraft and altering them to drop biodegradable cones filled with fertilizer and saplings over hard-to-access areas.[*][3]

32.2 See also

- *Miss Rumphius*, a 1982 children's book emphasizing public seed scattering

- Seed dispersal

- Johnny Appleseed

- Diggers

32.3 References

[1] Horton, Jennifer. "Could military strategy win the war on global warming?". How Stuff Works. Retrieved 2012-04-06.

[2] "How Guerrilla Gardening Works". How Stuff Works. Retrieved 2012-08-14.

[3] Brown, Paul (1999-09-02). "Aerial bombardment to reforest the earth". The Guardian. Retrieved 2011-06-09.

32.4 Further reading

- Smith, K. (2007). *The guerilla art kit*. Princeton Architectural Press.

- Huxta, B. (2009). *Garden-variety graffiti*. Organic gardening, 2009.

32.5 External links

- Stuffyoushouldknow.com

- Wikihow.com

- Latimes.com

- Gardenista.com

- Articles.washingtonpost.com

- Permanentculturenow.com

- News.bbc.co.uk

- Npr.org

- News.bbc.co.uk

- Theguardian.com

- UK Seed Bomb Supplier, Seed Freedom

- The guerrilla gardener's seedbomb recipe

- Planning an Effective Seed Bomb Strike

Chapter 33

SOS Mata Atlântica Foundation

The **SOS Mata Atlântica Foundation** was created in 1986 as a non-governmental and non-profit organization,[*][1] with the goal of defending what remains of Mata Atlântica (the Atlantic Forest) in Brazil.[*][2] Its actions are divided in six areas: public policies; campaigns; documentation, information and communication for conservation; environmental education and good citizenship; institutional development; and sustainable development, protection and handling of ecosystems.[*][3]

33.1 Projects

Its most notable project, ClickArvore,[*][4] aims at the reforestation of the Atlantic Forest. Since the year 2000,[*][5] it has planted about 22 million native trees, covering an area of approximately 130 km^2.[*][6] ClickArvore was idealized by Rodrigo Agostinho, who was elected mayor of Bauru in 2008.[*][7]

33.2 References

[1] Official Website of the SOS Mata Atlântica Foundation (in Portuguese)

[2] EarthDay's description of SOS Mata Atlântica Foundation as one of its partners

[3] Official Summarized Description of SOS Mata Atlântica in English

[4] Official Website of the ClickArvore project

[5] (Brazilian news) SOS Mata Atlântica lança site para reflorestamento, 06/08/2000. Folha de S.Paulo. http://www1.folha.uol.com.br/folha/informatica/ult124u871.shtml

[6] (Brazilian news) Clique uma árvore, 03/02/2011. Environmental News of the City of Itu. http://www.itu.com.br/meio-ambiente/noticia/clique-uma-arvore-20110203

[7] (Brazilian news) Prefeito eleito de Bauru tem 30 anos e diz que não vai apagar perfil do Orkut, 28/10/2008. Folha de S.Paulo. http://www1.folha.uol.com.br/folha/brasil/ult96u461097.shtml

Chapter 34

Tampa Bay Reforestation and Environmental Effort

Tampa Bay Reforestation and Environmental Effort, Inc. or more commonly known as "**T.R.E.E. Inc.**", is a 32-year-old grassroots nonprofit environmental organization based out of the Tampa Bay Area. It was instrumental in bringing the concept of volunteers raising and then planting trees along the interstates, roadways and in parks of the greater Tampa Bay Area to beautify and preserve the environment. To date, T.R.E.E. Inc. has planted over 26,397 trees at no cost to the taxpayer.

34.1 Early years 1983-1987

The organization started out as a small group of friends that wanted to see more trees being planted in the Tampa Bay Area. These friends took that purpose, moved forward on their own time and expense, and on February 8, 1983, T.R.E.E. Inc. was incorporated under Florida law.

T.R.E.E. Inc.'s *modus operandi* throughout most of its existence had been to purchase bare root tree seedlings, container grow them in 1-gallon containers for one growing season, step them up into 3-gallon containers for a second growing season and donate or out-plant them prior to their third growing season.

The variety of tree that was most commonly used during this period was the Genetically Improved or Superior North Florida Slash Pine (*Pinus elliottii var. "elliottii"*). It was selected due to its adaptability, rapid growth and relative ease of maintenance after establishment. Hardwood trees during that time were typically purchased as 4" potted seedlings. Varieties typically used were Sweetgum (*Liquidambar styraciflua*), Pignut Hickory (*Carya glabra*) and Loblolly Bay (*Gordonia lasianthus*).

34.2 Transition years 1988-1989

In January 1988, William Moriaty stepped down as President so that he could relocate to Gainesville, Florida with his wife Karen Cashon as she would be attending the University of Florida later that year. As a result, an almost entirely new slate of Directors served from 1988 to 1989. Vice President Bob Scheible was the only Founding member to serve during this two-year period in the same capacity that he did at the organization's creation.

34.3 Major program initiatives, 1990-2004

1990 was the major turning point that would set the stage for revised and newly created corporate philosophies that would guide T.R.E.E. Inc. for well over the next decade of its existence.

The Genetically Improved or Superior North Florida Slash Pine began to lose favor to the locally indigenous Longleaf Pine (*Pinus palustris*), "Ocala" Race Sand Pine (*Pinus clausa*) and South Florida Slash Pine (*Pinus elliottii var. "densa"*). The availability of Florida grown bare root hardwood seedlings further broadened the plant palette to include Baldcypress (*Taxodium distichum*), Tulip Poplar (*Liriodendron tulipifera*), and Southern Magnolia (*Magnolia grandiflora*) as well as dependable stand-bys, Sweetgum and Pignut Hickory.

Due to a major push by members living in Pinellas County, Florida, the organization's name was changed through Florida law on July 24, 1991 from its original title of Tampa Reforestation and Environmental Effort, Inc. to Tampa Bay Reforestation and Environmental Effort, Inc.

On August 9, 1991, T.R.E.E. Inc. was awarded 501 (c) (3) tax exempt status from the United States Internal Revenue

Service as an Educational organization.

34.4 Major sponsors and new programs 2005 to present

Beginning in 2005, T.R.E.E. Inc. secured its first major sponsor, Esurance. Esurance is an on-line insurance provider based out of San Francisco, California. Beginning with the *esurance St. Pete Beach Tree-Athalon* planting on September 24, 2005, T.R.E.E. Inc. has had an additional ten volunteer tree plantings sponsored by Esurance.

The Esurance plantings were a radical departure from T.R.E.E. Inc.'s previous plantings which were conducted almost solely through the use of 3-gallon material grown at its own nursery. This gave T.R.E.E. Inc. the unprecedented opportunity to plant large sized 30-gallon trees, affording it the opportunity to finally conduct plantings with a much higher visual impact. In addition, the Esurance projects led to a gradual phasing out of T.R.E.E. Inc. having to depend so heavily upon its own nursery to obtain trees.

In addition to the Esurance plantings, T.R.E.E. Inc. began introducing programs such as the *Tulip Poplar Repopulation Program, Orange and Seminole Counties, Florida* where East Central Florida Eco-Type Tulip Poplar (*Liriodendron tulipifera*) grown from seed of trees native to that area have been planted since that time in Orlando, Winter Park, Altamonte Springs, Sanford, and Casselberry, Florida.

Similar programs include the *Longleaf Pine Repopulation Program* in Temple Terrace, Florida and the *Egmont Key Reforestation Initiative* at Egmont Key State Park in Hillsborough County, Florida.

In December 2008, T.R.E.E. Inc. received a major contribution from The Home Depot Foundation and was also instrumental in assisting the N.F.L. Environmental Program and Florida Division of Forestry with a *Super Bowl Trail of Trees* planting initiative in April 2009.

On November 11, 2010, T.R.E.E. Inc. was instrumental in creating a listing of recommended flowering, conifer, hardwood and salt tolerant trees for the City of Dunedin's proposed Trailside Oasis Arboretum. On October 15, 2011, the first major installation of the Arboretum was made possible through the *Esurance Dunedin Trailside Oasis Arboretum Planting Project* that used a large proportion of low-chill temperate flowering trees such as "St. Lukes" Purple Leaf Plum, Taiwan Flowering Cherry and "Weaver" White Flowering Dogwood.

On October 15, 2012, T.R.E.E. Inc. participated in a "Scotties Trees Rock" planting project at Pepin Academies in Tampa.

On January 17, 2015, T.R.E.E. Inc. continued a decade old tradition of giving away free tree seedlings for the City of Temple Terrace's Florida Arbor Day Celebration. The organization has already secured tree seedlings for the City's 2016 Florida Arbor Day Celebration in January.

34.5 Vision 2020

The Vision 2020 Program was adopted on January 3, 2010 to be used as a guiding force in T.R.E.E. Inc.'s development and operations over the next ten years.

In addition to placing emphasis on continuing its Flowering Tree, Tulip Poplar, Longleaf Pine and South Florida Slash Pine reforestation initiatives, T.R.E.E. Inc. intends to expand operations into Manatee and Sarasota Counties, as well as convert all paper records into electronic form and register all corporate records with the United States Copyright Office of the Library of Congress. The first such electronic record to be forwarded to the United States Copyright Office was on August 15, 2010 featuring the organization's 1983 Annual Yearbook. Initially expected to take until the end of February 2011 to complete, all electronic registration of records dating from 1983 to 2010 were completed on February 11, 2011.

34.6 New Milestones

On July 14, 2012, the organization's 400th planting project occurred at the University of Florida Teaching Garden located at the Plant City, Florida campus of Hillsborough Community College. A 5-gallon Silver Maple (*Acer saccharinum*) was planted by Founding Member William Moriaty.

On February 9, 2013 the organization celebrated its 30th Anniversary by planting low-chill flowering trees at Serena Park in Temple Terrace, Florida. The highlight of the planting was a tree dedicated to the namesake of this Central Florida community, the "Temple" Tangor. At one time, Temple Terrace had one of the largest citrus groves in the world, consisting of 5,000 acres of this variety, which was itself named after the Founder of the Florida Citrus Exchange, William Chase Temple.

34.7 Mission

There are five major components to the organization's Mission:

1. Beautify and reforest the Tampa Bay area through the planting of native trees.

2. Further public awareness about the merits of reforesting public lands.

3. Further public awareness about the planting and preserving of native trees and underutilized trees of special interest.

4. Work in conjunction with, and in support of, other organizations similar purpose.

5. Be a clearinghouse of scientific and educational information to the public on the merits of tree species selection for use in the Tampa Bay area in accordance with the intent of its I.R.S. tax-exempt 501(c)(3) status.

34.8 Accomplishments

T.R.E.E. Inc. has over the past 32 years:

- Conducted 419 reforestation projects.

- Furnished the original design, and subsequent installation on November 23, 1996, of what would become the *Richard T. Bowers Historic Tree Grove* at the Museum of Science and Industry (Tampa).

- With the permission of the Florida Department of Transportation, made history by planting over 4,600 native trees in ten (10) *TREE DAY on I-75* projects along Interstate 75 in Hillsborough County, Florida from 1986 to 1996.

- Donated and planted approximately 26,397 native trees, 15,500 native wildflowers and 9,657 native shrubs, vines or ground covers utilizing an estimated 2,800 volunteers since T.R.E.E. Inc.'s creation on February 8, 1983.

- Assisted in a joint tree planting effort with the National Football League and Florida Division of Forestry consisting of over 20 projects in the Greater Tampa Bay Area for Super Bowl XLIII.

- On November 11, 2010, assisted Dunedin, Florida with a list of recommended trees for its proposed Trailside Oasis Arboretum, followed up with the Arboretum's first major planting on October 15, 2011.

- Assisted in the installation of a public community garden at Boyd Hill Nature Preserve in St. Petersburg, Florida beginning July 1, 2012.

- Celebrated its "Green Anniversary" by planting low-chill flowering trees and a "Temple" Tangor at Serena Park in Temple Terrace, Florida on February 9, 2013.

34.9 Newsletter

T.R.E.E. Inc. has a quarterly on-line newsletter available to the public called *Arbor Bio*. The newsletter can be found by linking to the external T.R.E.E. Web Site referenced below.

34.10 Notable directors and members

- Richard Strickland - President/Secretary 2012-2015 Secretary only, 2002-2012

- Bob Scheible - Founder and Vice President, 1983-2015

- John Blechschmidt - Treasurer, 1990-2003, 2009-2015

- Greg Howe - Founder and Secretary, 1983, 1985-1987 past President 1988-1989

- William Moriaty - Founder and past President, 1983-1987, 1990-2012

- Greg Van Stavern - Founder and past Treasurer, 1983 past Secretary 1984, 1990

- Nancy Buckley - Honorary Founder and past Treasurer, 1983-1987

- Kathy Caffentzis - Past Treasurer, 1988-1989

- Patricia Chellman - Past Secretary, 1988-1989

- Richard Bailey - Past Secretary, 1991-2001

- Jennifer Thompson - Past Public Relations Director, 2005-2006

- Jennifer Moore - Public Relations Coordinator, 2007-2015

- Dr. Sylvia Earle - Lifetime Member of T.R.E.E. Inc, past Chief Scientist of National Oceanic and Atmospheric Administration, presently "Explorer-in-Residence" with the National Geographic

34.11 Lifetime Members

- William Moriaty (Founder)

- Bob Scheible (Founder)

- Greg Van Stavern (Founder)

- Greg Howe (Founder)

- Autumn Balthazor
- Sheryl Bowman
- Cliff Brown
- Debbie Butts
- Dade City Garden Club
- Alice Earle (deceased)
- Dr. Sylvia Earle
- Ross J. Ferlita
- Hugh Gramling
- Horticultural Alliance, Inc.
- Harvey A. Hunt, P.E.
- William Jonson
- Brightman Logan
- Revello Medical Centers
- Rick Strickland
- Barbara Waddell (deceased)
- Dr. Richard Wunderlin
- Brett Youngster
- Richard A. Bailey (Honorary)
- John L. Blechschmidt Jr. (Honorary)
- Nancy Buckley (Honorary)
- Bob Der (Honorary)
- Kathy Caffentzis (Honorary)
- Steve Graham (Honorary)
- Gary L. Henry, R.L.A. (Honorary)
- Susana Thompson (Honorary)

34.12 Reference links

- T.R.E.E. Inc. Web Site

Chapter 35

Three-North Shelter Forest Program

"Green Great Wall" redirects here. It is not to be confused with the Great Green Wall, a similar anti-desertification effort in Saharan Africa.

The **Three-North Shelter Forest Program** (simplified Chinese: 三北防护林; traditional Chinese: 三北防護林; pinyin: *Sānběi Fánghùlín*), also known as the **Three-North Shelterbelt Program** or the **Green Great Wall**, is a series of human-planted windbreaking forest strips (shelterbelts) in China, designed to hold back the expansion of the Gobi Desert.[1] It is planned to be completed around 2050,[2] at which point it will be 2,800 miles (4,500 km) long.

The project's name indicates that it is to be carried out in all three of the northern regions: the North, the Northeast and the Northwest.[3]

35.1 Effects of the Gobi Desert

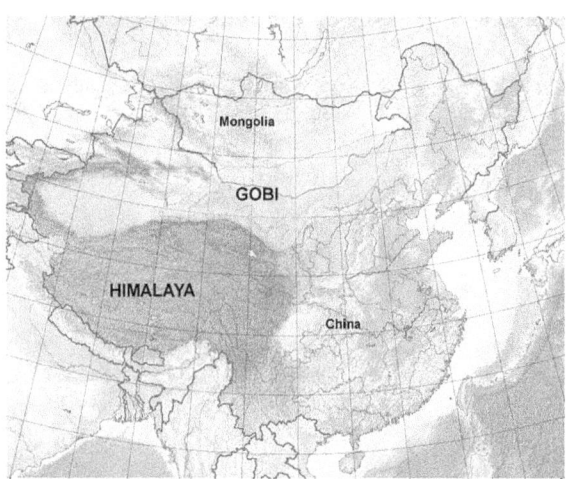

Map of China and the Gobi desert

China has seen 3,600 km^2 (1,400 sq mi) of grassland overtaken every year by the Gobi Desert.[4] Each year dust storms blow off as much as 2,000 km^2 (800 sq mi) of topsoil, and the storms are increasing in severity each year. These storms also have serious agricultural effects for other nearby countries, such as Japan, North Korea, and South Korea.[5] The Green Wall project was begun in 1978, with the proposed end result of raising northern China's forest cover from 5 to 15 percent [6] and thereby reducing desertification.

35.2 Methodology and progress

The fourth and most recent phase of the project, started in 2003, has two parts: the use of aerial seeding to cover wide swathes of land where the soil is less arid, and the offering of cash incentives to farmers to plant trees and shrubs in areas that are more arid.[7] A $1.2 billion oversight system (including mapping and surveillance databases) is also to be implemented.[7] The "wall" will have a belt with sand-tolerant vegetation arranged in checkerboard patterns in order to stabilize the sand dunes. A gravel platform will be next to the vegetation to hold down sand and encourage a soil crust to form.[7] The trees should also serve as a windbreak from dust storms.

35.3 Measuring success

As of 2009, China's planted forest covered more than 500,000 square kilometers (increasing tree cover from 12% to 18%) – the largest artificial forest in the world.[8] Of the 53,000 hectares planted that year, a quarter died.[6] In 2008 winter storms destroyed 10% of the new forest stock, causing the World Bank to advise China to focus more on quality rather than quantity in its stock species.[8]

35.4   Problems

If the trees succeed in taking root, they could soak up large amounts of groundwater, which would be extremely problematic for arid regions like northern China.[7] For example, in Minqin, an area in north-western China, studies showed that groundwater levels have dropped by 12–19 metres since the advent of the project.[6]

Land erosion and overfarming have halted planting in many areas of the project. China's booming pollution rate has also weakened the soil, causing it to be unusable in many areas.[4]

Furthermore, planting blocks of fast-growing trees reduces the biodiversity of forested areas, creating areas that are not suitable to plants and animals normally found in forests. "China plants more trees than the rest of the world combined," says John McKinnon, the head of the EU-China Biodiversity Programme. "But the trouble is they tend to be monoculture plantations. They are not places where birds want to live." The lack of diversity also makes the trees more susceptible to disease, as in 2000, when one billion poplar trees were lost to disease, setting back 20 years of planting efforts.[6]

Liu Tuo, head of the desertification control office in the state forestry administration, is of the opinion that there are huge gaps in the country's efforts to reclaim the land that has become desert.[9] At present there are around 1.73 million sq kilometers that have become desert in China, of which 530,000 km^2 are treatable. But at the present rate of treating 1,717 km^2 per year, it would take 300 years to reclaim the land that has become desert.[10]

35.5   Relations to climate change

China's forest scientists argue that monoculture tree plantations are more effective at absorbing the greenhouse gas carbon dioxide than slow-growth forests,[8] so while diversity may be lower, the trees purportedly help to offset China's carbon emissions. (See List of countries by carbon dioxide emissions)

35.6   Criticism

There are many who do not believe that the Green Wall is an appropriate solution to China's desertification problems. Gao Yuchuan, the Forest Bureau head of Jingbian County, Shanxi, stated that "planting for 10 years is not as good as enclosure for one year," referring to the alternative non-invasive restoration technique that fences off (en-

closes) a degraded area for two years to allow the land to restore itself.[6] Jiang Gaoming, an ecologist from the Chinese Academy of Sciences and proponent of enclosure, says that "planting trees in arid and semi-arid land violates [ecological] principles".[6] The worry is that the fragile land cannot support such massive, forced growth. Others worry that China is not doing enough on the social level. In order to succeed, many believe the government should encourage farmers financially to reduce livestock numbers or relocate away from arid areas.[7]

35.7   See also

- Buffer strip

- Energy-efficient landscaping

- Great Plains Shelterbelt, 1930s-40s, US

- Great Plan for the Transformation of Nature, 1940s-50s, Soviet Union

- Macro-engineering

- Sand fence

- Seawater greenhouse

- Deforestation and climate change

35.8   References

[1] "MEDIA REPORTS I China's Great Green Wall". BBC News. 3 March 2001-03-03. Retrieved 2012-05-19. Check date values in: |date= (help)

[2] "State Forestry Administration,P.R.China" (in Chinese). English.forestry.gov.cn. Retrieved 2012-05-19.

[3] 李谷城 (Li Kwok-sing) (2006). 中國大陸改革開放新詞語[A Glossary of New Political Terms of the PRC in the Post-Reform Era] (in Chinese). HK: Chinese University Press. p. 39. ISBN 978-962-996-258-6.

[4] "The Fall of the Green Wall of China". WorldChanging. 29 December 2003. Retrieved 17 March 2007.

[5] "China's Dust Storms Raise Fears of Impending Catastrophe". National Geographic. 1 June 2001. Retrieved 19 October 2009.

[6] "China's Great Green Wall Proves Hollow". The Epoch Times. 29 July 2009. Retrieved 19 October 2009.

[7] "The Green Wall Of China". Wired. April 2003. Retrieved 19 October 2009.

[8] Watts, Jonathan (11 March 2009). "China's loggers down chainsaws in attempt to regrow forests". London: The Guardian. Retrieved 19 October 2009.

[9] Jonathan Watts (4 January 2011). "China makes gain in battle against desertification but has long fight ahead | Environment". London: The Guardian. Retrieved 2012-05-19.

[10] Patience, Martin (2011-01-04). "BBC News - China official warns of 300-year desertification fight". Bbc.co.uk. Retrieved 2012-05-19.

35.9 External links

- China's Great Green Wall

- China's forest shelter project dubbed "green Great Wall"

- Grassland ecology to curb sandstorms

- Taming the Yellow Dragon - A Billion Trees in the Desert

- Taming the Yellow Dragon - The Korea Herald

- Verticall gardent in hanoi, vietnam

Chapter 36

Tree planting

A treeplanter in northern Ontario.

Tree planting is the process of transplanting tree seedlings, generally for forestry, land reclamation, or landscaping purposes. It differs from the transplantation of larger trees in arboriculture, and from the lower cost but slower and less reliable distribution of tree seeds.

In silviculture the activity is known as reforestation, or afforestation, depending on whether the area being planted has or has not recently been forested. It involves planting seedlings over an area of land where the forest has been harvested or damaged by fire or disease or insects. Tree planting is carried out in many different parts of the world, and strategies may differ widely across nations and regions and among individual reforestation companies. Tree planting is grounded in forest science, and if performed prop-

Tree planting is an aspect of habitat conservation. In each plastic tube a hardwood tree has been planted.

erly can result in the successful regeneration of a deforested area. Reforestation is the commercial logging industry's answer to the large-scale destruction of old growth forests, but a planted forest rarely replicates the biodiversity and complexity of a natural forest.

Because trees remove carbon dioxide from the air as they grow, tree planting can be used as a geoengineering technique to remove CO_2 from the atmosphere.

36.1 By country

36.1.1 Australia

Australian forests have been heavily affected since European colonisation, and some attempts have been made to restore native habitats, both by government and individuals. Greening Australia is a national Non profit set up to run the "National Tree Program" initiated by the Federal Government in 1982.[1] Greening Australia completed the 1 Billion Tree target and has gone on to become one of the major tree planting organisations in the country. There is a strong volunteer movement for conservation in Australia through Landcare*[2] and other networks. National Tree Day is organised annually by Planet Ark in the last week in July, encouraging the public to plant 1 million native trees per year. Many state governments run their own "Million Tree" programs each year to encourage community involvement.*[3]*[4]

36.1.2 Canada

Most tree planting in Canada is carried out by private reforestation companies.*[5] Reforestation companies compete with one another for contracts from logging companies, whose annual allowable cut for the following year is based upon how much money they invest into reforestation and other silvicultural practices. Treeplanting is typically piece work and tree prices can vary widely depending on the difficulty of the terrain and on the winning contract's bid price. As a result, there is a saying among planters: "There is no bad land, only bad contracts." 4 months of hard work can yield enough to live on for an entire year, but conditions are brutal.*[5]

Tree planting crews often do not permanently reside in the areas where they work, thus much planting is based out of motels or bush camps. Bush camp accommodations usually consist of a mess tent, cook shack, dry goods tent, first aid tent, freshly dug outhouses, and a shower tent or trailer. Planters are responsible for bringing either a tent or car to sleep in. A camp also contains camp cooks and support staff.*[5]

Planting is carried out in accordance to the client's specifications, and planters are expected to learn the quality standards for each contract that they work on. Planted clearcuts are spot checked on a regular basis. Although quality concerns vary across contracts, spot checkers are typically looking for such things as: species appropriate site choice, species appropriate spacing, how tight the seedlings are in the ground, how straight the seedlings are, and whether or not the seedlings have been damaged. These concerns vary from region to region, and from contract to contract.

The average British Columbian planter plants 1 600 trees per day,*[6] but it is not uncommon for veterans to plant up to 4,000 trees per day while working in the interior.*[5] These numbers are higher in central and eastern Canada, where the terrain is generally faster, however the price per tree is slightly lower as a result. Average daily totals of 2500 are common, with experienced planters planting upwards of 5000 trees a day. Numbers as high as 7500 a day have been recorded.*[5] Planters typically work 8–10 hours per day with an additional 1 to 2 hours of (usually) unpaid traveling time. Work weeks on British Columbian planting contracts

are usually 4–5 days long, with 1–2 days off. In Ontario, work weeks are generally 5–6 days long, with 1 day off.

Quite often, tree planting contractors will deduct some of the cost associated with the operation of the contract directly from the tree planter's daily earned wages. These imposed fees typically vary from $10 to $30 per day, and are referred to as "camp costs" . In *some* cases, rookie tree planters end up owing their employer money for the first few pay periods.*[7]

Once inflation is factored in, real tree planter earnings have declined for many years in Canada. This has adversely affected the sector's ability to attract and retain workers.*[8] Higher wages and much better working conditions in many other industries, from construction, to oil and gas, and even information technology, has led to fewer Canadian young people wanting to plant trees.

Based on statistics for British Columbia, the average tree planter: lifts a cumulative weight of over 1,000 kilograms (2,200 lb), bends more than 200 times per hour, drives the shovel into the ground more than 200 times per hour and travels over 16 kilometres (9.9 mi) with a heavy load, every day of the entire season. The reforestation industry has an average annual injury rate of approximately 22 claims per 100 workers, per year. It is often difficult and sometimes dangerous.*[6]

36.1.3 Great Britain

Planting in Britain is commonly referred to as *restocking*, when it takes place on land that has recently been harvested. When occurring on previously unforested land it is known as *new planting*.*[9] Under the British system, in order to acquire the necessary permissions to clearcut, the landowner must agree a management plan with the Forestry Commission (the regulatory body for all things forestry) which must include proposals for the re-establishment of tree cover on the land. Planting contractors will be engaged by the landowner/management company, a contract drawn up and work will typically take place from November to April when most of the transplants are dormant.

Planting is part of the rotational nature of much British plantation forestry. Productive tree crops are planted and subsequently clearcut. Some form of soil cultivation may take place and the ground is then restocked. Where the production of timber is a management priority, a prescribed stocking density must be achieved. For coniferous species this will be a minimum of 2500 stems per hectare at year 5 (from planting). Planting at this density has been shown to favour the development of straighter knot-free logs.

Planters are normally paid under piece work terms and an experienced worker will plant around 1500 trees a day under most conditions.

36.1.4 Israel

Tree-planting is an ancient Jewish tradition. The Talmudic rabbi Yohanan ben Zakai used to say that if a person planting a tree heard that the Messiah had arrived, he should finish planting before going to greet him.[10] With over 240 million planted trees, Israel is one of only two countries that entered the 21st century with a net gain in the number of trees. Due to massive afforestation efforts,[11] this fact echoed in diverse campaigns.[12][13] Israeli forests are the product of a major afforestation campaign by the Jewish National Fund (JNF).[14]

The largest planted forest in Israel is Yatir Forest, located on the southern slopes of Mount Hebron, on the edge of the Negev Desert. It covers an area of 30,000 dunams (30 square kilometers).[15] It is named after the ancient Levite city within its territory, Yatir, as written in the Torah: "And unto the children of Aaron the priest they gave Hebron with its suburbs, the city of refuge for the manslayer, and Libnah with its suburbs, and Jattir with its suburbs, and Eshtemoa with its suburbs" (Book of Joshua 21:13-14).[16] In 2006, the JNF signed a 49-year lease agreement with the State of Israel which gives it control over 30,000 hectares of Negev land for the development of forests.[17] Research on climate change is being carried out in Yatir Forest.[18][19] Studies of the Weizmann Institute of Science, in collaboration with the Desert Research Institute at Sde Boker, have shown that the trees function as a trap for carbon in the air.[20][21] Shade provided by trees planted in the desert also reduces evaporation of the sparse rainfall.[20] Yatir Forest is a part of the NASA project FluxNet, a global network of micrometeorological tower sites used to measure the exchanges of carbon dioxide, water vapor, and energy between terrestrial ecosystem and atmosphere. The Arava Institute for Environmental Studies conducts research that focuses on crops such as dates and grapes grown in the vicinity of Yatir forest.[22][23] The research is part of a project aimed at introducing new crops into arid and saline zones.[24]

The JNF has been criticized for planting non-native pine trees which are unsuited to the climate, rather than local species such as olive trees.[25] Others say that JNF deserves credit for this decision, and the forests would not have survived otherwise.[26] According to JNF statistics, six out of every 10 saplings planted at a JNF site in Jerusalem do not survive, although the survival rate for planting sites outside Jerusalem is much higher – close to 95 percent. Critics argue that many JNF lands outside the West Bank were illegally confiscated from Palestinian refugees, and that the JNF furthermore should not be in-

volved with lands in the West Bank.[27] Shaul Ephraim Cohen has claimed that trees have been planted to restrict Bedouin herding.[28] Susan Nathan wrote that forests were planted on the site of abandoned Arab villages after the 1948 war.[29] Nathan also writes that olive trees were replaced by pine and cypress trees[30] and that JNF afforestation policy erases traces of the Arab presence prior to 1948.[31]

Since 2009, the JNF has provided the Palestinian Authority with 3,000 tree seedlings for a forested area being developed on the edge of the new city of Rawabi, north of Ramallah.[32]

Approximately one thousand small forest fires are registered on average every year during the five fire-prone months. Half of them are caused by arson, hostile actions and Arab or Palestinian terrorist attacks. Ten thousand acres of hand-planted forest were destroyed by Katyusha rockets during the 2006 Lebanon War by Hezbollah. In summer 2006, JNF launched Operation Northern Renewal, a reforestation effort, which also replaced some topsoil that was burned away.[33]

36.1.5 New Zealand

Kaingaroa Forest in New Zealand is the largest planted forest in the southern hemisphere. It is one of the many plantation forests planted since European settlement. The Monterey Pine (*Pinus radiata*) is commonly used for plantations since a fast-growing cultivar suitable for a wide range of conditions has been developed.

Government agencies, environmental organisations and private trusts carry out tree planting for conservation and climate change mitigation. While some work is carried out by private enterprise there are also planting days organised for volunteers. Landcare Research use planted forests for their EBEX21 system for greenhouse gas emissions mitigations.[34]

36.1.6 South Africa

South Africa's forests have been heavily depleted mostly due to agriculture, traditional farming and urbanisation in the coastal regions, various organizations are working on increasing the forest cover in parts of the country. Currently there is less than 0.5% forest cover in South Africa. Greenpop is a national Social Enterprise set up to run a "Tree Planting Program" greening both urban and rural areas. This was initiated in 2010. There is a strong volunteer movement for conservation in South Africa. National Tree Day or Arbor Day is organised annually in September, and has gone on to become national Arbor Month.

36.1.7 United States

Trees for the Future and Plant With Purpose are non-profit organizations based in the U.S. that plant trees in developing countries to improve land management.*[35]*[36] Other organizations that plant trees in the United States include:

- American Forests

- Arbor Day Foundation*[37]

- Nature Conservancy

- Plant-it 2020*[38]

- USDA Forest Service "Plant-A-Tree" program in which a person can donate to plant trees in the National Forests.*[39]

- Our City Forest*[40]

- TreeFolks empowers central Texans to build stronger communities through planting and caring for trees. Since 1989, TreeFolks has planted over 1.5 million trees in parks, neighborhoods, and natural areas throughout central Texas.

36.2 Role in climate change

The development of markets for tradeable pollution permits in recent years have opened up a new source of funding for tree planting projects: carbon offsets. The creation of carbon offsets from tree planting projects hinges on the notion that trees help to mitigate climate change by sequestering carbon dioxide as they grow. However, the science linking trees and climate change is largely unsettled, and trees remain a controversial source of offsets.

36.2.1 Climate impacts

Climate scientists working for the IPCC believe human-induced global deforestation is responsible for 18-25% of global climate change. The United Nations, World Bank and other leading nongovernmental organizations are encouraging tree planting to mitigate the effects of climate change.

Trees sequester carbon through photosynthesis, converting carbon dioxide and water into molecular dioxygen (O_2) and plant organic matter, such as carbohydrates (e.g., cellulose). Hence, forests that grow in area or density and thus increase in organic biomass will reduce atmospheric CO_2 levels. (Carbon is released as CO_2 if a tree or its lumber burns or decays, but as long as the forest is able to grow back at the same rate as its biomass is lost due to oxidation of organic carbon, the net result is carbon neutral.) In their 2001 assessment, the IPCC estimated the potential of biological mitigation options (mainly tree planting) is on the order of 100 Gigatonnes of carbon (cumulative) by 2050, equivalent to about 10% to 20% of projected fossil fuel emissions during that period.*[41]

However, the global cooling effect of forests from carbon sequestration is not the only factor to be considered. For example, the planting of new forests may initially release some of the area's existing carbon stores into the atmosphere. Specifically, the conversion of peat bogs into oil palm plantations has made Indonesia the world's third largest producer of greenhouse gases.*[42]

Compared to less vegetated lands, forests affect climate in three main ways:

- Cooling the Earth by functioning as carbon sinks, and adding water vapor to the atmosphere and thereby increasing cloudiness.

- Warming the Earth by absorbing a high percentage of sunlight due to the low reflectivity of a forest's dark surfaces. This warming effect, or reduced albedo, is large where evergreen forests, which have very low reflectivity, shade snow cover, which is highly reflective.

To date, most tree planting offsets strategies have taken only the first effect into account. A study published in December 2005 combined all these effects and found that tropical forestation has a large net cooling effect, because of increased cloudiness and because of high tropical growth and carbon sequestration rates.*[43]

Trees grow three times faster in the tropics than in temperate zones; each tree in the rainy tropics removes about 22 kilograms (50 pounds) of carbon dioxide from the atmosphere each year.*[44] However, this study found little to no net global cooling from tree planting in temperate climates, where warming due to sunlight absorption by trees counteracts the global cooling effect of carbon sequestration. Furthermore, this study confirmed earlier findings that reforestation of colder regions —where long periods of snow cover, evergreen trees, and slow sequestration rates prevail —probably results in global warming. According to Ken Caldeira, a study co-author from the Carnegie Institution for Science, "To plant forests outside of the tropics to mitigate climate change is a waste of time." .*[45]

His premise that grassland reflects more sun, keeping temperatures lower, is, however, applicable only in arid regions. A well-watered lawn, for example, is as green as a tree, but absorbs far less CO_2. Deciduous trees also have the advantage of providing shade in the summer and sunlight in the winter; so these trees, when planted close to houses, can be utilized to help increase energy efficiency of these houses.

This study remains controversial and criticized for assuming dark colored trees might replace the frozen, white tundra in the upper northern hemisphere. Regular tree planting projects typically take place on lands that are only slightly different in color. The warming impact was also measured over hundreds of years, rather than a 30-70 year time horizon most climate experts believe we have to fix climate change.

Furthermore, the described warming effect (of temperate and boreal latitude forest) is only apparent once the trees have grown to create a dense 'close canopy', and it is at precisely this point that trees grown for offset purposes should be harvested and their absorbed carbon fixed for the long-term as timber.

36.2.2 Costs

While the benefits of tree planting are subject to debate, the costs are low[46] compared to many other mitigation options. The IPCC has concluded that "The mitigation costs through forestry can be quite modest (US$0.1–US$20 / metric ton carbon dioxide) in some tropical developing countries.... The costs of biological mitigation, therefore, are low compared to those of many other alternative measures" .[41] The cost effectiveness of tropical reforestation is due not only to growth rate, but also to farmers from tropical developing countries who voluntarily plant and nurture tree species which can improve the productivity of their lands.[47] As little as US$90 will plant 900 trees, enough to annually remove as much carbon dioxide as is annually generated by the fossil-fuel usage of an average United States resident.

36.2.3 Types of trees planted

The type of tree planted may have great influence on the environmental outcomes. It is often much more profitable to outside interests to plant fast-growing species, such as eucalyptus, casuarina or pine (e.g., *Pinus radiata* or *Pinus caribaea*), even though the environmental and biodiversity benefits of such monoculture plantations are not comparable to native forest, and such offset projects are frequently objects of controversy.

To promote the growth of native ecosystems, many environmentalists advocate only indigenous trees be planted. A practical solution is to plant tough, fast-growing native tree species which begin rebuilding the land. Planting non-invasive trees that assist in the natural return of indigenous species is called "assisted natural regenera-

A eucalyptus plantation in final stages at Arimalam.

tion." There are many such species that can be planted, of which about 12 are in widespread use, such as *Leucaena leucocephala*.[48] Alternatively, farmer-managed natural regeneration (FMNR), involves farmers preserving trees (not replanting), and is considered to be a more cost effective method of reforestation than regular tree planting.

36.3 See also

- Arbor Day

- Billion Tree Campaign

- Farmer-managed natural regeneration

- Hoedad (tool)

- Hoedads Reforestation Cooperative

- Mattock

- Planet Ark

- Plant-for-the-Planet

- Pottiputki

- Tu Bishvat

- Tubestock

- Urban reforestation

36.4 References

[1] "Greening Australia - History" .

[2] "Landcare" .

[3] "2 Million Trees Victoria".

[4] "Growing A Great Future". *SA Urban Forests Million Trees Program*. Government of South Australia. Retrieved December 13, 2013.

[5] Brittany Shoot (December 18, 2011). "The Dark Side of Reforestation Programs: Planting 7,000 Trees a Day in Brutal Conditions". *AterNet*. Retrieved December 26, 2011.

[6] "Preventing Tree Planting Injuries" (PDF). *Work Safe BC*. Workers' Compensation Board of British Columbia. 2006. Retrieved December 13, 2013.

[7] Archived November 20, 2008 at the Wayback Machine

[8] Betts, John (2007-07-30). "2007 Planting Season: More Planters—Less Experience". *Current Affairs*. Western Silvicultural Contractors' Association. Retrieved 2010-11-12.

[9] "Forestry Statistics 2005". *Forestry Commission: Economics and Statistics*. Forestry Commission. 2005. Retrieved December 13, 2013.

[10] "President of German States Council of Education Ministers Plants Tree at Kennedy Memorial". *Jerusalem Post* (The Jerusalem Post). July 29, 2009. Retrieved December 13, 2013. |section= ignored (help)

[11] "Israel Forestry & Ecology". Jewish National Fund, East 69th Street, NY 10021 USA. Retrieved 29 October 2011.

[12] "Trees from Israel" (PDF). standwithus.com. Archived from the original (PDF) on November 17, 2006. Retrieved 29 October 2011.

[13] "Five Widely-Read Bloggers Tour Israel and Plant Trees". standwithus.com. Retrieved 29 October 2011.

[14] "JNF Tree Planting Center". Jewish National Fund, East 69th Street, NY 10021, USA. Retrieved 29 October 2011.

[15] "Planting of Yatir Forest". Fr.jpost.com. 2013-12-17. Retrieved 2013-12-21.

[16] "JPost | French-language news from Israel, the Middle East & the Jewish World". Fr.jpost.com. 2013-12-17. Retrieved 2013-12-21.

[17] Professor Alon Tal, The Mitrani Department of Desert Ecology, The Blaustein Institutes for Desert Research, Ben Gurion University of the Negev. "NATIONAL REPORT OF ISRAEL, Years 2003-2005, TO THE UNITED NATIONS CONVENTION TO COMBAT DESERTIFICATION (UNCCD)"; State of Israel, July 2006

[18] Sahney, S., Benton, M.J. & Falcon-Lang, H.J. (2010). "Rainforest collapse triggered Pennsylvanian tetrapod diversification in Euramerica" (PDF). *Geology* **38** (12): 1079–1082. doi:10.1130/G31182.1.

[19] Bachelet, D; R.Neilson,J.M.Lenihan,R.J.Drapek (2001). "Climate Change Effects on Vegetation Distribution and Carbon Budget in the United States" (PDF). *Ecosystems* **4** (3): 164–185. doi:10.1007/s10021-001-0002-7.

[20] Issar, Arie (2009-11-30). "Benefits of planting trees in the desert". Haaretz.com. Retrieved 2013-12-21.

[21] KKL-JNF cooperating on afforestation at Yatir forest

[22] Vu du Ciel-documentary by Yann Arthus-Bertrand

[23] "2000 year old seed grows in the arava". Watsonblogs.org. Retrieved 2013-12-21.

[24] MERC Project M-20-0-18 project

[25] Rabbi David Seidenberg."The Giving Tree: A Way to Honor Our Vision for Israel"; Neohasid, 2006

[26] "JPost | French-language news from Israel, the Middle East & the Jewish World". Fr.jpost.com. 2013-12-17. Retrieved 2013-12-21.

[27] Dan Leon."The Jewish National Fund: How the Land Was 'Redeemed': The JNF's historical concept of exclusively Jewish land is wholly anachronistic"; Palestine-Israel Journal, Vol 12 No. 4 & Vol 13 No. 1, 05/06

[28] Shaul Ephraim Cohen. "The Politics of Planting"; University of Chicago 1993 p.121

[29] Nathan, Susan (2005). The Other Side of Israel: My Journey Across the Jewish/Arab Divide. New York: Nan A. Talese. pp. 130–131. ISBN 978-0-385-51456-9.

[30] Nathan, Susan (2005) op cit pages 129–130

[31] Nathan, Susan (2005) op cit pages 151–152

[32] Gross, Tom (2009-12-02). "Building Peace Without Obama's Interference". Online.wsj.com. Retrieved 2013-12-21.

[33] "Forestry & Ecology". Retrieved 29 October 2011.

[34] EBEX21, Carbon Credits System

[35] "Trees for the Future". Plant-trees.org. Retrieved 2013-12-21.

[36] "Plant With Purpose". Plant With Purpose. Retrieved 2013-12-21.

[37] "Replanting," Arbor Day Foundation.

[38] Plant-it 2020

[39] "Plant-A-Tree" program, USDA Forest Service

[40] Our City Forest

[41] Working Group III (July 2001). Bert Metz, ed. *Climate Change 2001: Mitigation. Third Annual Report*. World Meteorological Organization (Intergovernmental Panel on Climate Change). doi:10.2277/0521015022. Retrieved 2007.

[42] Delft Hydraulics (2006-07-12). "PEAT-CO_2: Assessment of CO_2 emissions from drained peat lands in SE Asia." (PDF). Wetlands International. Retrieved 2007.

[43] S. G. Gibbard; K. Caldeira; G. Bala; T. Phillips; M. Wickett; Lawrence Livermore National Laboratory; Carnegie Institution of Washington (2005-10-29). "Climate effects of global land cover change". *Geophysical Research Letters* **32**: L23705. Bibcode:2005GeoRL..3223705G. doi:10.1029/2005GL024550. Retrieved 2007-02-22.

[44] "Global Cooling Centers". Trees for the Future. 2006. Retrieved 2007.

[45] Jha, Alok (2006-12-15). "Planting trees to save planet is pointless, say ecologists". The Guardian. To plant forests to mitigate climate change outside of the tropics is a waste of time

[46]

[47] "Providing farmers and communities in the tropics with long-term assistance implementing environmentally and economically sustainable technologies". Sustainable Harvest International. Archived from the original on September 27, 2009. Retrieved 2007.

[48] Dave Deppner; John Leary; Karin Vermilye; Steve McCrea (2005). *The Global Cooling Answer Book* (PDF) (Second ed.). Trees for the Future. ISBN 1-879857-20-0. Retrieved 2007.

36.5 External links

- - Canadian research on treeplanting injury prevention

- Silviculture Magazine - free digital magazine about all things silviculture/reforestation

- Photography - Documentary Photography of Canadian Treeplanting

- Replant - Non-commercial Canadian site with thousands of planting photos

- Hard-Core Tree Planters - Advice from a tree planting veteran

- Tree-Planter.com - Information on tree planting as a job and community tips

- Most Tree Plantation in 24 Hour in Multiple Location World Records in India.

- Trailer for *78 Days*, Tree Planting Documentary

Chapter 37

Tree shelter

Young trees sheltered by plastic tubes

A **Tree shelter**, or **tree guard**, is a type of plastic shelter used to nurture trees in the early stages of their growth. Tree shelters are also sometimes known as **Tuley tubes** or tree tubes.

The purpose of tree shelters is to protect young trees from browsing by herbivores by forming a physical barrier along with providing a barrier to chemical spray applications. Additionally, tree tubes accelerate growth by providing a mini-greenhouse environment that reduces moisture stress, channels growth into the main stem and roots and allows efficient control of weeds that can rob young seedlings of soil moisture and sunlight.

Tree shelters were invented in the United Kingdom in 1979 by Graham Tuley (Lantagne 1997). They are particularly popular in the UK in landscape-scale planting schemes and their use has been established in the United States since 2000. About 1 million shelters were in use in the United Kingdom in 1983–1984 (Tuley 1985), and 10 million were produced in 1991 (Potter 1991).

Many variations of tree shelters exist. There is considerable debate among tree shelter manufacturers as to the ideal colour, size, shape and texture for optimal plant growth. One style used in northern climates of North America has a height of 5 feet to offer the best protection from deer browse, with vent holes in the upper portion of the tube to allow for hardening off of hardwood trees going into the winter months and no vent holes in the lower portion to shield seedlings from herbicide spray and rodent damage.

37.1 Economics

Tree shelters must be assessed alongside alternatives such as fencing to keep animals out, loss if no shelter is employed and visual impacts.

37.2 References

- Lantagne, D. O. (1997). "Using Tree Shelters To Establish Northern Red Oak And Other Hardwoods" (PDF). *Michigan State University Extension Bulletin*: 2. Retrieved 5 April 2015.

- Potter, M. J. (1991) *Treeshelters - Forestry Commission Handbook 7* HMSO

- Tuley, G. (1985). "The growth of young oak trees in shelters". *Forestry* **58** (2): 181–19. doi:10.1093/forestry/58.2.181.

37.3 External links

Chapter 38

Trees for the Future

Trees for the Future is a Maryland-based nonprofit organization founded in 1989 that helps communities around the world plant trees. Through seed distribution, agroforestry training, and in-country technical assistance, it has empowered rural groups to restore tree cover to their lands, protect the environment and help to preserve traditional livelihoods and cultures for generations.

38.1 Overview

Started in 1989 by Grace and Dave Deppner, Trees for the Future works with communities in Central America, South America, Africa and Asia to incorporate tree planting into their agricultural activities.*[1]

A 501(c)(3) nonprofit based in Silver Spring, Maryland, Trees for the Future also offers individuals and businesses a form of carbon offset through planting trees.*[2]

Its programs help communities replenish their natural resources by providing materials and training to allow farmers to sustainably grow crops for food, fodder, and fuelwood.*[3]

Trees for the Future has worked on reforestation efforts in Haiti*[4] since 2002.*[5] It is currently planting fast-growing Moringa trees which can help restore degraded farmlands.*[6] Trees for the Future's Haiti Coordinator, Timote Georges, was featured on Discovery Channel about the organization's work in Haiti working on agroforestry projects to restore degraded land throughout the country.*[7]

In 2008, Trees for the Future helped over 200 farmers plant approximately 250,000 forest and fruit trees in Léogâne, the epicenter of the earthquake, and worked with other rural communities. In 2009, due to the organization's increasing network of people on the ground and organizations supporting its work, the Trees for the Future's program had even more success. The organization's Haiti Coordinator, Timote Georges, continues to work with communities along the Arcadine coast to plant trees.*[8]*[9]

The organization to date has planted over 65 million trees worldwide in 30 countries and has served over 11,000 villages around the world. Trees for the Future provides free distance and agroforestry training and education; works in conjunction with over 53 specialists who are experts in agroforestry, community development, sustainable agriculture, land use, livestock management, women in development and youth education; provides in-country seed distribution, and; works on natural resource management.

Trees for the Future created a documentary about its work planting trees called "50 Million Trees and Counting: Trees for the Future".*[10]

In February 2010, Maryland Senators Richard Madaleno and Brian Frosh, and Maryland Delegate Al Carr announced Senate resolutions recognizing the 20th anniversary of Trees for the Future and the organization's two decades of global activity restoring degraded lands and cutting carbon emissions through the planting of more than 65 million trees.*[11]

Trees for the Future partnered with SodaStream International in 2012 to launch the Replant Our Planet initiative. Sodastream committed to planting ten trees in Brazil for each home beverage carbonation system sold from its Rethink Your Soda product line.*[12]

38.2 See also

38.3 References

[1] "Trees for the Future". Charity Navigator. Accessed April 15, 2010.

[2] Underwood, Kristin. "Trees For the Future: Don't Count Your Footprint, Just Plant a Tree". *Treehugger*. July 21, 2009.

[3] Hayden, Erik. "Tree by Tree: Reforesting Haiti". *Miller-McCune* magazine. January 27, 2010.

[4] Phillips, Jessi. "Interview: Trees for the Future of Haiti". *Sierra Club:The Green Life.* April 1, 2010.

[5] LaFranchi, Howard. "Haiti's pressing need: rain-resistant shelter for 750,000 homeless". *Christian Science Monitor.* February 9, 2010.

[6] Tomassini, Jason. "Silver Spring nonprofit planting trees in Haiti". *Maryland Gazette.* April 7, 2010.

[7] The Discovery Channel. "EchoHeroes:Timote Georges". *Discovery Channel Global Education.* March 4, 2010.

[8] Lewis, Sunny. "Haiti's Few Trees At Risk as Survivors Flee to Rural Areas". *Environment News Service.* January 26, 2010.

[9] Interlandi, Jeneen. "A Tree Grows in Haiti". *Newsweek.* July 16, 2010.

[10] Deppner, Dave. "50 Million Trees and Counting: Trees for the Future". YouTube. November 30, 2006.

[11] General Assembly of Maryland "Resolution Trees for the Future". Senate of Maryland: 2010 Regular Session. February 23, 2010.

[12] Munarriz, Rick Aristotle (22 March 2012). "SodaStream Wants You to Hug a Tree, Drink a Soda". *The Motley Fool.* Retrieved 29 March 2012.

38.4 External links

- Trees for the Future official Web site
- Trees for the Future Facebook Page

Chapter 39

Xanthosoma sagittifolium

Xanthosoma sagittifolium, the **arrowleaf elephant ear** or **arrowleaf elephant's ear**, is a species of tropical flowering plant in the genus *Xanthosoma*, which produces an edible, starchy corm.

39.1 Cuisine and reforestation

In Bolivia, it is called *walusa*, in Colombia *bore*, in Costa Rica *tiquizque* or *macal*, in Mexico *mafafa*, in Nicaragua *quequisque*, and in Panama *otoy*. In Brazil, the leaves are sold as *taioba*. The tuber (called *nampi* or *malanga*) is also used in the cuisine of these countries. The plant is often interplanted within reforestation areas to control weeds and provide shade during the early stages of growth.

In Puerto Rican cuisine, the plant and its corm are called *yautia*. In Puerto Rican *pasteles*, *yautia* is ground with squash, potato, green bananas and plantains into a dough-like fluid paste containing pork and ham, and boiled in a banana leaf or paper wrapper. The *yautia* corm is used in stews, soups, or simply served boiled much like a potato. It is used in local dishes such as *guanime*, *alcapurrias*, *sancocho*, and *mondongo*. In *alcapurrias*, it is also ground with green bananas and made into fried croquettes containing *picadillo* or sea food. *Yautia majada* is also prepared and consumed when mashed in some instances. *Yauita* puree is usually served with fish or shell fish cooked in coconut milk.

In Suriname and the Netherlands, the plant is called *tayer*. The shredded root is baked with chicken, fruit juices, salted meat, and spices in the popular Surinamese dish, *pom*. Eaten over rice or on bread, *pom* is commonly eaten in Suriname at family gatherings and on special occasions, and is also popular throughout the Netherlands.

39.2 References

[1] "The Plant List" .

[2] "Tropicos.org" .

39.3 External links

- GRIN Species Profile - *Xanthosoma sagittifolium*

- FAO Species Profile: Xanthosoma sagittifolium

- Costa Rica Ministry of Agriculture and Livestock: Cultivation manual (Spanish)

39.4 Text and image sources, contributors, and licenses

39.4.1 Text

- **Reforestation** *Source:* https://en.wikipedia.org/wiki/Reforestation?oldid=683155161 *Contributors:* William Avery, Jaknouse, Infrogmation, Aarchiba, WhisperToMe, SEWilco, Robbot, Bkell, Alan Liefting, MPF, Mackeriv, Sam Hocevar, CALR, Discospinster, Vsmith, Joel Russ, Guettarda, Vortexrealm, Jeberlie, 119, Ziegenpeter, Belirac, Padraic, Rjwilmsi, Vegaswikian, KAM, KFP, Chobot, WriterHound, Wavelength, Okedem, Gaius Cornelius, Neum, RazorICE, Ravedave, Epipelagic, Antoshi, Silverchemist, Closedmouth, Arthur Rubin, HereToHelp, Meegs, Carlosguitar, SmackBot, Amcbride, Big Adamsky, KVDP, Eskimbot, Bluebot, Ottawakismet, SchfiftyThree, Bondfool, Beilby, Rama's Arrow, Can't sleep, clown will eat me, Rrburke, GVnayR, Cybercobra, Zzorse, Michael Rogers, Byelf2007, Valfontis, Rigadoun, Longshot14, JoeBot, IvanLanin, Tawkerbot2, JForget, Cydebot, Peripitus, Chasingsol, Balamche, Alaibot, Finn krogstad, TruthbringerToronto, HappyInGeneral, S Marshall, Inner Earth, AntiVandalBot, PatMcClendon, Jj137, Ingolfson, MikeLynch, JAnDbot, Acroterion, Connormah, Canjth, Pawl Kennedy, DancingPenguin, Artaxiad, J.delanoy, The Boy that time forgot, Ncmvocalist, Fincaproject, Max stolkin, Oscargsol, DASonnenfeld, JJGD, Mark v1.0, Shellyblake, Oxfordwang, Lamro, AlleborgoBot, Earthtree, Logan, Rstafursky, Nopetro, Steven Crossin, Gift Of Ireland, Ravanacker, OK-Bot, FOBRHAWT, Rvannatta, ClueBot, Roppo, Mild Bill Hiccup, Niceguyedc, Takeaway, Downtowngal, Bryan.Seigneur, Lujar, HveyVermnt, Maky, Ost316, Avoided, Addbot, Cxz111, CurtisSwain, AkhtaBot, Download, Chamal N, SpBot, Saimiriwildlife~enwiki, Tassedethe, Woolalah, Jarble, Legobot, Yobot, AnomieBOT, Wisg, Mygreenspace, Krasss, Minnecologies, Brunhildr, Citation bot, Reader454, Sylwia Ufnalska, Anna Frodesiak, Ita140188, Bellerophon, Brambleshire, Carrite, Amaury, Adyione, Mpritza, Thehelpfulbot, Pinethicket, Triplestop, Cavallero, The Ent, Digitat, TheStrayCat, Dinamik-bot, MegaSloth, RjwilmsiBot, TjBot, Slon02, Mukogodo, John of Reading, Oliverlyc, Look2See1, Wikipelli, Dhammadarsa Bhikkhu, Joshgreene, Nijoint, ClueBot NG, Tabletrack, Widr, Helpful Pixie Bot, Gob Lofa, BG19bot, Northamerica1000, Mark Arsten, Wikiman897, Anigelseth, G.Kiruthikan, Mark viking, JimnChina, LazyReader, Monkbot, Isabellahicks, Meerp1, Marcel Hendrik, Lavhenia, Cailynjhkim and Anonymous: 188

- **Urban reforestation** *Source:* https://en.wikipedia.org/wiki/Urban_reforestation?oldid=672061725 *Contributors:* Alan Liefting, Bender235, Diego Moya, Wavelength, Jim.henderson, DASonnenfeld, Pjoef, SwisterTwister, AnomieBOT, Jonesey95, Trappist the monk, Northamerica1000, Mark Arsten, Mogism, Emilybbrodie, Mark viking and Anonymous: 1

- **Afforestation** *Source:* https://en.wikipedia.org/wiki/Afforestation?oldid=685621113 *Contributors:* SimonP, Ike9898, SEWilco, Pgan002, Vsmith, ParticleMan, Polylerus, Jaardon, Belirac, Woohookitty, WadeSimMiser, Daniel Collins, KAM, Krueschan, Guliolopez, Vmenkov, Wavelength, Alynna Kasmira, Rjensen, Epipelagic, KVDP, Agentbla, Tomtefarbror, Jtm71, Eliyak, Tymek, Robofish, Optakeover, André Koehne, Ziusudra, ChrisCork, MarsRover, Cydebot, Igby, S Marshall, Gökhan, The Transhumanist, Twsx, Engineman, Ronbarak, Kiore, CommonsDelinker, Sankars098, DASonnenfeld, Kimdime, Idioma-bot, Funandtrvl, Philip Trueman, Bentogoa, Flyer22, Lightmouse, Gift Of Ireland, Denisarona, ClueBot, Wordup 10, Cenarium, Herunar, WikiHead, Badgernet, Cewvero, Addbot, Favonian, Jarble, Yobot, Hohenloh, AnomieBOT, Piano non troppo, NSK Nikolaos S. Karastathis, Sylwia Ufnalska, Anna Frodesiak, Brutaldeluxe, Eddaoddsdottir, FrescoBot, Dmitriy01, Pinethicket, Elockid, Winayak1410, Tea with toast, Dcmaster33, Look2See1, Racerx11, Salsero35, Donner60, Ego White Tray, 28bot, Petrb, ClueBot NG, Lhb1239, Tcarignan, Adityamadhav83, MerllwBot, Helpful Pixie Bot, Varenilus, BG19bot, Ramesh Ramaiah, Northamerica1000, Solomon7968, Rokowsky, Suchthekaitlin, TheJJJunk, حموده عزمي داراي, G.Kiruthikan, Habibibibalani, Nikhil.gupta.jgd, Bugg t, LithiumEnergy, Ugog Nizdast, Ginsuloft, Cypherquest, RADfamily, DarkestElephant, Lizia7, Stamptrader, LazyReader, Olenyash, Santi-Lak, Parveetsid, Cailynjhkim, GeneralizationsAreBad and Anonymous: 129

- **Afforestation in Japan** *Source:* https://en.wikipedia.org/wiki/Afforestation_in_Japan?oldid=666061989 *Contributors:* Wavelength, Rjensen, DASonnenfeld, Boneyard90, AnomieBOT, Northamerica1000, Khazar2, Everymorning, Zapechacek, Brownbarons, Cailynjhkim, Sedicesimo and Anonymous: 1

- **Arbor Day** *Source:* https://en.wikipedia.org/wiki/Arbor_Day?oldid=685087009 *Contributors:* Bryan Derksen, Olivier, Earth, Ixfd64, Zani-mum, Tregoweth, Docu, Angela, Darkwind, Kaihsu, Jeandré du Toit, Lukobe, Mxn, Adam Bishop, דוד, Wetman, Branddobbe, Moncrief, Nyh, Academic Challenger, Bkell, Alan Liefting, Wwoods, MacGyverMagic, PFHLai, Tellumo, Demiurge, Canterbury Tail, Warfieldian, D6, Stepp-Wulf, Discospinster, Rich Farmbrough, Paul August, ESkog, Swid, Bobo192, Smalljim, Dpaajones, Markmusante, Pearle, Jonathunder, Alan-sohn, Arthena, DreamGuy, Wtmitchell, Jon Cates, Cmapm, Kazvorpal, Weyes, MONGO, Acerperi, Tabletop, Anthony Dean, Arzachel, Hard Raspy Sci, Zzyzx11, Graham87, Rjwilmsi, Tim!, PinchasC, Bubba73, JYOuyang, Celestianpower, MentalTypo, Bgwhite, Cornellrockey, Wave-length, TexasAndroid, Brabo~enwiki, Lexicon, Zwobot, Marcelin0, Asarelah, Evrik, Pete Simpson, Tim1965, Crunch, Rogue 9, SmackBot, Djravery, Tom Lougheed, Brick Thrower, Lsommerer, Gilliam, Fnorth, Baa, A. B., DavidSpencer.ca, Duncancumming, Addshore, Oanabay04, Noles1984, Lisasmall, Risssa, The alliance, The Frederick, Bjankuloski06, Bella Swan, Tjs2012, Robotchoir, B7T, FactChecker, Joseph Solis in Australia, Jeffrey3732, Kylu, Melicans, Cydebot, Mjaquez, Future Perfect at Sunrise, Tedcoombs, Gogo Dodo, Lugnuts, Luccas, Avi4now, Ame-liorate!, Billmason, BetacommandBot, Epbr123, Mojo Hand, Benqish, Mentifisto, AntiVandalBot, Majorly, T L Miles, Dricherby, LordFoom, VoABot II, Glen, DerHexer, Littlepear, MartinBot, T1Rex, CommonsDelinker, Johnpacklambert, DandyDan2007, SimpsonDG, Liangent, Rochelimit, Bailo26, Arms & Hearts, Ndunruh, Ahuskay, Vanished user 39948282, Feats-O-Strength, DASonnenfeld, Xiahou, VolkovBot, ABF, Naveen.nswamy, Barneca, Gzdavidwong, Hypnopomp, Imasleepviking, Martin451, Chengdi, Zenswashbuckler, ChristopherCashell, Sevela.p, Victor.wang, Garywolf777, ISAYsorry, Tomalak geretkal, Dtreed, Happysailor, Lightmouse, Ted Duke, Gunmetal Angel, Johnnywiggle, Wiki-Laurent, Superbeecat, Xnatedawgx, Tatterfly, MBK004, ClueBot, Voxpuppet, EoGuy, FieldMarine, Gonei72, Niceguyedc, Djloopwrex, Excirial, Cayambe, Taifarious1, Lartoven, Arjayay, SoxBot III, Egmontaz, DumZiBoT, XLinkBot, ItsLassieTime, Spitfire, AMRDeuce, Dthomsen8, MystBot, MMich, Addbot, Captain-tucker, Choctaw06, Douglas the Comeback Kid, Ka Faraq Gatri, Samsonjackson, Klingercrazy, Einstein-greco, Favonian, AtheWeatherman, Dayofswords, Arxiloxos, Yobot, AnomieBOT, Essin, Dwayne, Crecy99, Materialscientist, Ingenosa, E235, Optimusprimechucknorris, Obersachsebot, 4twenty42o, Natracha, DSisyphBot, Myoon7891, Wwbread, Marcelivan, Brutaldeluxe, Shadow-jams, Thehelpfulbot, FrescoBot, Rubenescio, SaturnineMind, Oldlaptop321, Laleksa, Bdinchitwn, Newcellphone, D'ohBot, Doremo, Audioli-brary, HamburgerRadio, Pinethicket, I dream of horses, Onthegogo, VenomousConcept, Full-date unlinking bot, Blodance, Lotje, Dinamik-bot, TBloemink, Minimac, Onel5969, NerdyScienceDude, Mderowitsch, DASHBot, Gfoley4, 1234567yo, Sheeana, Trickytruck, ZéroBot, Zbase4, John Cline, Daonguyen95, Ὁ οἶστρος, Aaron90omar, Mcmatter, Erianna, Seducationofficer, Rcsprinter123, Jay-Sebastos, Ollllllllll-lie, Howl56, Orange Suede Sofa, Hasol753, M4pnt, Yrtimid, Spicemix, ClueBot NG, Mattosoft, MelbourneStar, Joefromrandb, Plmoknqwerty,

O.Koslowski, Widr, Chillllls, Ryan Vesey, IgnorantArmies, Nebraskan11, Mightymights, Rhbsihvi, Gaspar19500, BG19bot, Kndimov, Rich-levine00, MusikAnimal, Scottman95, Tutelary, PortalandPortal2Rocks, Pratyya Ghosh, ChrisGualtieri, Padenton, JYBot, Andtrace, EagerTod-dler39, Xochiztli, Webclient101, 331dot, Lugia2453, Eyesnore, Cmckain14, Armandpaz2, Hitdarbway, Taxus2000, Finealt, Iman90, Puuuuj, Najaska, Sfahleson, Shex4lby, Lesliejenky, ToonLucas22 and Anonymous: 356

- **Armenia Tree Project** *Source:* https://en.wikipedia.org/wiki/Armenia_Tree_Project?oldid=650650534 *Contributors:* Edward, Lquilter, Kral-izec!, Wavelength, Badagnani, StephenWeber, Arthur Rubin, Serouj, Cydebot, Prolog, Clariosophic, DASonnenfeld, VartanM, Solar-Wind, Folklore1, Minnecologies, FrescoBot, Mogism, Kahtar, Juhuyuta, Coltnicastro and Anonymous: 9

- **The Big Tree Plant** *Source:* https://en.wikipedia.org/wiki/The_Big_Tree_Plant?oldid=674211260 *Contributors:* Tim!, The Anomebot2, DA-Sonnenfeld, Deor and Ivolocy

- **Billion Tree Campaign** *Source:* https://en.wikipedia.org/wiki/Billion_Tree_Campaign?oldid=678582454 *Contributors:* Bearcat, Alan Liefting, Nard the Bard, BanyanTree, Pol098, Rjwilmsi, Tim!, Membender, SmackBot, Shalom Yechiel, Rrburke, Ohconfucius, Dl2000, CommonsDelinker, Katharineamy, Skullers, DASonnenfeld, PNG crusade bot, AlleborgoBot, Resurgent insurgent, Billion Tree Campaign - UNEP, Dani75, Redthoreau, Good Olfactory, Addbot, AnomieBOT, Rubinbot, Materialscientist, FrescoBot, Dinamik-bot, H3llBot, Rebecca Louise Carter, CarbonFreeDining, Greppo, ClueBot NG, Karanshah46, Ehr1Ros2, Coolworlds12345, TheMGB, Monicagellar 08 and Anonymous: 16

- **Biosequestration** *Source:* https://en.wikipedia.org/wiki/Biosequestration?oldid=677299645 *Contributors:* Nealmcb, William M. Connolley, Alan Liefting, StuartH, Vsmith, YUL89YYZ, Rjwilmsi, Arthur Rubin, SmackBot, RDBrown, Mion, Shirifan, Gobonobo, Shane204, ChrisCork, Cydebot, Christian75, Bhaskarmv, Id447, DuncanHill, Beagel, R'n'B, GirasoleDE, Bettis211, WereSpielChequers, Niceguyedc, Auntof6, Aw-ickert, Addbot, Debresser, Bermicourt, Citation bot, Eumeng.chong, FrescoBot, Nepomuk 3, Citation bot 1, Jkw0010, Crusoe8181, Nim-busWeb, Merlinsorca, Leaf82, Mendo23, ZéroBot, Helpful Pixie Bot, Bibcode Bot, Jkw0020, Matulkar, JuvenisUrsus, Danny Sprinkle, S Gord B, Perla.Marroquin, Monkbot and Anonymous: 33

- **Community forests in England** *Source:* https://en.wikipedia.org/wiki/Community_forests_in_England?oldid=558701348 *Contributors:* MRSC, Tim!, Oliver Chettle, Zzuuzz, Mais oui!, SmackBot, Verne Equinox, Choalbaton, Simply south, Rettetast, DASonnenfeld, Matthew Brandon Yeager, Ethantait, Denisarona, Mattgirling, Yobot, Crookesmoor, Northamerica1000, Illia Connell, Uisto and Anonymous: 3

- **Farmer-managed natural regeneration** *Source:* https://en.wikipedia.org/wiki/Farmer-managed_natural_regeneration?oldid=675607754 *Contributors:* Skysmith, Alan Liefting, RJFJR, RHaworth, SmackBot, JFHJr, Chris the speller, Shrumster, Stwalkerster, Iridescent, Alai-bot, Kateshortforbob, Johnbod, Marynz, S, DASonnenfeld, Hugo999, Bearian, Paul.dettmann, Tony.rinaudo, Kernel Saunters, Editore99, Lu Wunsch-Rolshoven, Kbdankbot, Addbot, Lightbot, Yobot, In2thats12, Look2See1, Dewritech, ZéroBot, ClueBot NG, BG19bot, Okailey, Tim-lam92, Timlovesflowers, Seandixonsul and Anonymous: 5

- **Forest landscape restoration** *Source:* https://en.wikipedia.org/wiki/Forest_landscape_restoration?oldid=615980945 *Contributors:* Guettarda, Nthep, Magioladitis, DASonnenfeld, Dolphin51, SwisterTwister, Forru and Anonymous: 1

- **Forest restoration** *Source:* https://en.wikipedia.org/wiki/Forest_restoration?oldid=650080262 *Contributors:* Guettarda, Wavelength, DASon-nenfeld, FrescoBot, OgreBot, Werieth, Mogism, Forru, Lizia7, Christian7robert and Anonymous: 5

- **Forests for the 21st Century** *Source:* https://en.wikipedia.org/wiki/Forests_for_the_21st_Century?oldid=666651104 *Contributors:* Lquilter, Alan Liefting, Bender235, Mandarax, JonHarder, Ser Amantio di Nicolao, SMasters, Shawn in Montreal, DASonnenfeld, Bovineboy2008, Gueneverey, Addbot, Yobot, Fortdj33, WQUlrich, Banej, Cmmorton, TheHappiestCritic and DoctorKubla

- **Futuro Forestal S.A.** *Source:* https://en.wikipedia.org/wiki/Futuro_Forestal_S.A.?oldid=666088168 *Contributors:* Bearcat, Wavelength, DA-Sonnenfeld, Tassedethe, LilHelpa, Frietjes, BG19bot and Arnofk

- **Gap dynamics** *Source:* https://en.wikipedia.org/wiki/Gap_dynamics?oldid=642734916 *Contributors:* Rjwilmsi, CombatWombat42, Ateth-nekos, Yobot, MrX, RjwilmsiBot, Ego White Tray, Lugia2453, KSmith1105, Etwillia, Monkbot, Nmcke1 and Anonymous: 3

- **Green Belt Movement** *Source:* https://en.wikipedia.org/wiki/Green_Belt_Movement?oldid=676497341 *Contributors:* Sjc, Shizhao, Pengo, Alan Liefting, Bobblewik, Addicted, Vsmith, Nard the Bard, Stbalbach, Thaths, DaeX, Tibetibet, FlaBot, Wavelength, Thane, Jonathan.s.kt, SmackBot, JFHJr, Gilliam, Hectorguinness, Julius Sahara, Flickety, Ohconfucius, Cydebot, Hebrides, Luna Santin, Callumgrieve, Pax: Vobiscum, R.Schuster, DorganBot, DASonnenfeld, VolkovBot, Iwillseetheworld, Mimihitam, ClueBot, PipepBot, Cmbook2, L.tak, Frood, Addbot, Numbo3-bot, Lightbot, Yobot, Fraggle81, A Stop at Willoughby, Jim1138, ליטו+ר ', Xqbot, Cordeliab, Thehelpfulbot, FrescoBot, Mayraunc, K6ka, Rcsprinter123, LZ6387, Dexbot, Riyandarson, KasparBot, Xovady and Anonymous: 37

- **Groasis Waterboxx** *Source:* https://en.wikipedia.org/wiki/Groasis_Waterboxx?oldid=672723668 *Contributors:* Gidonb, Discospinster, Vsmith, BD2412, TDogg310, Eastmain, KylieTastic, DASonnenfeld, Wikiisawesome, SchreiberBike, XLinkBot, Addbot, Luckas-bot, Yobot, The Banner, GrouchoBot, FrescoBot, LucienBOT, I dream of horses, Horst-schlaemma, TGCP, WikitanvirBot, ChuispastonBot, BG19bot, Northamerica1000, Gorthian, HueSatLum, Ingeev and Anonymous: 11

- **Hauberg** *Source:* https://en.wikipedia.org/wiki/Hauberg?oldid=678041884 *Contributors:* Bermicourt and YiFeiBot

- **Hoedads Reforestation Cooperative** *Source:* https://en.wikipedia.org/wiki/Hoedads_Reforestation_Cooperative?oldid=647149842 *Contrib-utors:* Lquilter, Anarchivist, Tedder, SONORAMA, SmackBot, Valfontis, Gobonobo, Aboutmovies, DASonnenfeld, Yobot, FrescoBot, Ham-burgerRadio, Look2See1, Helpful Pixie Bot, BattyBot, Gk4johnson and Anonymous: 2

- **International Small Group and Tree Planting Program** *Source:* https://en.wikipedia.org/wiki/International_Small_Group_and_Tree_Planting_Program?oldid=621244822 *Contributors:* Brianhe, Malcolma, SmackBot, Magioladitis, Belovedfreak, DASonnenfeld, Hugo999, Phil Bridger, Addbot, Unrev, Erik9bot, Skyerise, Look2See1, Helpful Pixie Bot, BG19bot and Treeplanterske

- **Kidney tray (tool)** *Source:* https://en.wikipedia.org/wiki/Kidney_tray_(tool)?oldid=586873619 *Contributors:* Dennis Brown, DASonnenfeld, Addbot, Wikiman897, Nerdyboy6057 and GrandDaSarge

- **Theodore Lukens** *Source:* https://en.wikipedia.org/wiki/Theodore_Lukens?oldid=679477765 *Contributors:* Choster, Bearcat, Giraffedata, Woohookitty, BD2412, Rjwilmsi, Jivecat, Wavelength, RussBot, Michael Slone, Attilios, Neo-Jay, Amakuru, Ewulp, BeenAroundAWhile, Hekerui, KudzuVine, DASonnenfeld, Squids and Chips, Copana2002, Lightmouse, Mild Bill Hiccup, Marcia Wright, Download, Samhuddy, AnomieBOT, Eumolpo, LilHelpa, FrescoBot, PigFlu Oink, Cnwilliams, John of Reading, Look2See1, Helpful Pixie Bot, VIAFbot, Junk-yardsparkle, LaurentianShield, MLynneK, KasparBot, Knife-in-the-drawer, Srednuas Lenoroc and Anonymous: 4

- **Million Tree Initiative** *Source:* https://en.wikipedia.org/wiki/Million_Tree_Initiative?oldid=617691711 *Contributors:* Alan Liefting, Ground Zero, ShelfSkewed, Alaibot, L'Aquatique, DASonnenfeld, DerBorg, DumZiBoT, Staticshakedown, Aleaiactaest07, Addbot, Fluffernutter, Yobot, Tohd8BohaithuGh1, LilHelpa, Merlinsorca and Anonymous: 2

- **Monterey County reforestation** *Source:* https://en.wikipedia.org/wiki/Monterey_County_reforestation?oldid=549479536 *Contributors:* Woohookitty, RHaworth, Malcolma, SmackBot, Wizardman, Iridescent, CmdrObot, Cydebot, Alaibot, Jllm06, The Anomebot2, Emeraude, Lightmouse, Addbot, Rscote, Lightbot, Yobot, Muzknut, Mean as custard, Look2See1 and Anonymous: 5

- **Mount Airy Forest** *Source:* https://en.wikipedia.org/wiki/Mount_Airy_Forest?oldid=678975057 *Contributors:* Mxn, Chris Capoccia, Doncram, Gilliam, Hmains, Greg5030, Jllm06, Od Mishehu AWB, Look2See1, BattyBot, ChrisGualtieri, Monkbot and Anonymous: 3

- **The National Forest (England)** *Source:* https://en.wikipedia.org/wiki/The_National_Forest_(England)?oldid=686703675 *Contributors:* Rmhermen, Renata, Ronz, Jschwa1, Reddi, Wellington, Pigsonthewing, Bobblewik, Jrdioko, Andycjp, Keith Edkins, Mrtrey99, Grunners, Noisy, Chris j wood, Cap, Naturenet, Pperos, Grutness, Nik42, Joolz, Saga City, Pcpcpc, Mindmatrix, Awostrack, JBellis, Pszomszor, Kbdank71, Tim!, Oliver Chettle, Kanthoney, Conscious, Neilbeach, Ravedave, Light current, GEWJ, Mais oui!, Smurfy, Mhkay, Jhartshorn, Bluebot, Charivari, Regan123, Dumelow, Dl2000, Tawkerbot2, Trident13, RobotG, Michig, Frankie816, Jllm06, J.delanoy, Victuallers, Amgmichael, Ginga123, Relocaterock, DASonnenfeld, GrahamHardy, BotMultichill, Sjwells53, Lightmouse, ImageRemovalBot, AdeleBeeby, 718 Bot, Fidodogsimmons2, Mhockey, Lightbot, Yobot, AnomieBOT, SVDTorvaldsSKS, Kwiki, NMAguide, Gigogag, A930913, Ivolocy, RobinLeicester, Costesseyboy, BlueStar303, National Forest, Cdwn, Lommes and Anonymous: 23

- **NZ Native Forests Restoration Trust** *Source:* https://en.wikipedia.org/wiki/NZ_Native_Forests_Restoration_Trust?oldid=654184884 *Contributors:* DavidLevinson, Alan Liefting, MPF, Bobblewik, Cavrdg, Man vyi, Grutness, Bleakcomb, Tom Webb, Dialectric, SmackBot, Robofish, Cydebot, DASonnenfeld, L.tak, Good Olfactory, AnomieBOT, Minnecologies, Schwede66 and Nfrt1

- **Pottiputki (tool)** *Source:* https://en.wikipedia.org/wiki/Pottiputki_(tool)?oldid=662279662 *Contributors:* DocWatson42, Wavelength, Rwalker, Just plain Bill, Amartyabag, Sobreira, Lfstevens, DASonnenfeld, SchreiberBike, Addbot, Jylöstalo, Rcsprinter123, BattyBot, Wikiman897, NahidSultan, Sminthopsis84, Andrei Marzan, Lch29, Brandonbecktxst, Positron77, Maphead354, Jokerswild117, Killa luigi and Anonymous: 2

- **Reducing emissions from deforestation and forest degradation** *Source:* https://en.wikipedia.org/wiki/Reducing_emissions_from_deforestation_and_forest_degradation?oldid=679478257 *Contributors:* Kizor, Bender235, Rjwilmsi, KVDP, Dl2000, Faizhaider, DASonnenfeld, Yone Fernandes, Correogsk, Jarble, AnomieBOT, RjwilmsiBot, John of Reading, Winner 42, ThePowerofX, Petrb, KLBot2, Valoradam, BattyBot, Dexbot, GSDPhilip, Monkbot, Pvanlaake, Ges neiu and Anonymous: 11

- **Secondary forest** *Source:* https://en.wikipedia.org/wiki/Secondary_forest?oldid=680270518 *Contributors:* Fred Bauder, 0x6D667061, Neutrality, Kbh3rd, Guettarda, Bobo192, Man vyi, Grutness, Vadim Makarov, Tabletop, GregorB, Cataclysm, KAM, Hmains, Snowmanradio, Anlace, Accurizer, Cydebot, Thijs!bot, Ufwuct, Joan-of-arc, Alphachimpbot, Nyttend, Bobanny, McSly, DASonnenfeld, Wowter, Xuweiyi1987~enwiki, Addbot, Darkwishster, Jarble, Luckas-bot, Andrewrp, Minnecologies, Xqbot, Anna Frodesiak, Brambleshire, Joostik, Sanzoneja, Tiramisoo, Fod2009, EmausBot, Look2See1, Glacialfox, Quackriot and Anonymous: 23

- **Seed bombing** *Source:* https://en.wikipedia.org/wiki/Seed_bombing?oldid=663267327 *Contributors:* BenFrantzDale, JarlaxleArtemis, Wavelength, Stikman, Rwalker, SmackBot, KVDP, Cacuija, Bluebot, Thumperward, Derek R Bullamore, Smallfri, Kayobee, JustAGal, Widefox, Seaphoto, Brianmanden, Donloggins, Hysocc, Greywaterbrown, Addbot, JBsupreme, Other of us, EmausBot, Gbadg, Look2See1, AvicAWB, 1Veertje, Urbanlandscout, ChrisGualtieri, Iberichard, Bananasoldier, Eyesnore, Flynniekins1125 and Anonymous: 26

- **SOS Mata Atlântica Foundation** *Source:* https://en.wikipedia.org/wiki/SOS_Mata_Atl%C3%A2ntica_Foundation?oldid=685469075 *Contributors:* Lquilter, Bearcat, Hmains, Chris the speller, Victor Lopes, Katharineamy, Funandtrvl, Ceilican, Addbot and Yobot

- **Tampa Bay Reforestation and Environmental Effort** *Source:* https://en.wikipedia.org/wiki/Tampa_Bay_Reforestation_and_Environmental_Effort?oldid=683243958 *Contributors:* Angilbas, Alan Liefting, SoWhy, Woohookitty, BD2412, Wavelength, Colonies Chris, Majorclanger, Jllm06, WOSlinker, Flyer22, Wuhwuzdat, Airplaneman, Hairhorn, Eumolpo, Fortdj33, Zaxby, JBofTampabay, Pennsuco, Look2See1, Candleabracadabra, Fraulein451, Khazar2, American Money, Treeguy55, DeliriousTreeman and Anonymous: 11

- **Three-North Shelter Forest Program** *Source:* https://en.wikipedia.org/wiki/Three-North_Shelter_Forest_Program?oldid=676891370 *Contributors:* The Anome, Menchi, Selket, Shizhao, Quadalpha, Alan Liefting, Markussep, Bobo192, Anthony Appleyard, Alai, Chobot, Vmenkov, Eraserhead1, Pigman, Robyvecchio, Arthur Rubin, SmackBot, Hmains, CSWarren, Rigadoun, 041744, Joseph Solis in Australia, Teratornis, JuWiki, Stefan Jensen, Ilovetuna, The Anomebot2, Engineman, Cgingold, Chauncey freak, Balthazarduju, DASonnenfeld, Hugo999, TXiKiBoT, Karmos, Piperh, Naive rm, Crash Underride, JuWiki2, Lightmouse, ClueBot, Coinmanj, 7, Dana boomer, Tdslk, WikHead, Airplaneman, Addbot, Lightbot, Wikimono111, Poko, Apothecia, Cureden, Ita140188, Thehelpfulbot, LucienBOT, Pinethicket, Elekhh, ZhBot, RjwilmsiBot, EmausBot, ECTaiwan2010, ZéroBot, Bxj, MRant71295, ClueBot NG, Michaelmas1957, Gob Lofa, Cold Season, Vandhana297, BattyBot, Acadēmica Orientālis, ChrisGualtieri, Monkbot, Cailynjhkim and Anonymous: 58

- **Tree planting** *Source:* https://en.wikipedia.org/wiki/Tree_planting?oldid=686194041 *Contributors:* Fred Bauder, G-Man, Aarchiba, Denni, Nv8200pa, Jeffq, Mattflaschen, Alan Liefting, Jpgordon, Guettarda, Jeberlie, Craigy144, Wtmitchell, Duff, Mandarax, Padraic, Miq, Rjwilmsi, Bgwhite, Jeremy Visser, SmackBot, Amcbride, Tarret, Durova, Chris the speller, OrphanBot, Gobonobo, Dombackpack, Gilabrand, Vanisaac, ChrisCork, Crazekid, Raysonho, Funnyfarmofdoom, Foofish, Hippypink, Toohool, Ingolfson, Gomm, Mmacdo97, Pnels081, The Boy that time forgot, Javndyke, KylieTastic, WJBscribe, DASonnenfeld, SieBot, Andrewjlockley, Uwaga, Nopetro, CutOffTies, Huggi, Denisarona, Excirial, Ottre, Sansumaria, A3camero, BOTarate, XLinkBot, Tommichaluk, Mythriam, Dthomsen8, Tubedog, AffinityQuence, Addbot, DOI bot, GreenSarah, AkhtaBot, Leszek Jańczuk, Earthmonger, Profundity loaf, Make comment, Ws227, TinyHelmsman, HerculeBot, Legobot, Luckas-bot, Yobot, Fraggle81, Hughstimson, Minnecologies, Citation bot, Ronewirl, Anna Frodesiak, J04n, The Interior, Brutaldeluxe, FrescoBot, Menwith, Dogposter, Pinethicket, Nimmonet, Tea with toast, Stalwart111, Balablitz, Look2See1, Rcsprinter123, Graphicpictures, The Infantree, ClueBot NG, Widr, Karanshah46, Wbm1058, BG19bot, The Banner Turbo, Northamerica1000, Vjf555, StinGer56, KaterM, Wikiman897, Cyberbot II, ChrisGualtieri, Mark vining, Chinaman68, Jodosma, Cunningchrisw, Datdyat, Babitaarora, Monkbot, Soloism, Chriswalker85, Abegnale, Pisicutapispispis, Faheemsalam and Anonymous: 114

- **Tree shelter** *Source:* https://en.wikipedia.org/wiki/Tree_shelter?oldid=672024709 *Contributors:* JoJan, Rich Farmbrough, Bdk, SP-KP, Rjwilmsi, MacRusgail, Gurch, Mais oui!, Chris the speller, Dlohcierekim, The Boy that time forgot, DASonnenfeld, Fadesga, Rainbowpro, Dawynn, Minnecologies, Citation bot, EQR~enwiki, Anna Frodesiak, Pinethicket, EmausBot, WikitanvirBot, Look2See1, ClueBot NG, Forester66, Stinkfart101, Whizz40 and Anonymous: 10

- **Trees for the Future** *Source:* https://en.wikipedia.org/wiki/Trees_for_the_Future?oldid=673647850 *Contributors:* Lquilter, Smalljim, Wavelength, Dialectric, SmackBot, KVDP, Colonies Chris, Magioladitis, Gorav, Jllm06, DASonnenfeld, Aymatth2, Solar-Wind, XLinkBot, Yobot, Pinethicket, Whywhenwhohow, GoingBatty, Ajbeckwith, Doorknobbish, ClueBot NG, Biosketch, Josvebot, Cameron0207, Inphynite and Anonymous: 4

- **Xanthosoma sagittifolium** *Source:* https://en.wikipedia.org/wiki/Xanthosoma_sagittifolium?oldid=674599833 *Contributors:* Hesperian, Richard Arthur Norton (1958-), Badagnani, IceCreamAntisocial, EncycloPetey, Takowl, Cydebot, Richard New Forest, Jaguarlaser, Brewcrewer, Chhe, Addbot, Flakinho, Luckas-bot, Xqbot, Gigemag76, Liberia88, FrescoBot, Fortdj33, GaryD1945, TobeBot, Boricuamark, EmausBot, Look2See1, Plantdrew, Sminthopsis84, MARCELIN DE BADJOB and Anonymous: 9

39.4.2 Images

- **File:12_year_old_restoration_plot_Doi_Suthep-Pui_National_Park_N._Thailand.jpg** *Source:* https://upload.wikimedia.org/wikipedia/commons/8/88/12_year_old_restoration_plot_Doi_Suthep-Pui_National_Park_N._Thailand.jpg *License:* CC BY-SA 3.0 *Contributors:* Self-photographed (Original text: *I took the photo*) *Original artist:* Forru

- **File:1_year_after_planting_framework_tree_species.jpg** *Source:* https://upload.wikimedia.org/wikipedia/commons/c/c7/1_year_after_planting_framework_tree_species.jpg *License:* CC BY-SA 3.0 *Contributors:* Photographed during field monitoring *Original artist:* Forru

- **File:2009ArborDayRochesterMinnesota.JPG** *Source:* https://upload.wikimedia.org/wikipedia/commons/9/92/2009ArborDayRochesterMinnesota.JPG *License:* CC BY-SA 3.0 *Contributors:* Transferred from en.wikipedia; transferred to Commons by User:Logan using CommonsHelper.
Original artist: Jonathunder (talk). Original uploader was Jonathunder at en.wikipedia

- **File:Aegopodium_podagraria1_ies.jpg** *Source:* https://upload.wikimedia.org/wikipedia/commons/b/bf/Aegopodium_podagraria1_ies.jpg *License:* CC-BY-SA-3.0 *Contributors:* Own work *Original artist:* Frank Vincentz

- **File:Ambox_current_red.svg** *Source:* https://upload.wikimedia.org/wikipedia/commons/9/98/Ambox_current_red.svg *License:* CC0 *Contributors:* self-made, inspired by Gnome globe current event.svg, using Information icon3.svg and Earth clip art.svg *Original artist:* Vipersnake151, penubag, Tkgd2007 (clock)

- **File:Ambox_important.svg** *Source:* https://upload.wikimedia.org/wikipedia/commons/b/b4/Ambox_important.svg *License:* Public domain *Contributors:* Own work, based off of Image:Ambox scales.svg *Original artist:* Dsmurat (talk · contribs)

- **File:Ambox_wikify.svg** *Source:* https://upload.wikimedia.org/wikipedia/commons/e/e1/Ambox_wikify.svg *License:* Public domain *Contributors:* Own work *Original artist:* penubag

- **File:Arbor_day_in_Algeria.jpg** *Source:* https://upload.wikimedia.org/wikipedia/commons/2/29/Arbor_day_in_Algeria.jpg *License:* CC BY 2.0 *Contributors:* http://www.flickr.com/photos/mekfouldji/5113969440/sizes/o/in/photostream/ *Original artist:* amekinfo

- **File:Arisaema_triphyllum_fruit.jpg** *Source:* https://upload.wikimedia.org/wikipedia/commons/f/f4/Arisaema_triphyllum_fruit.jpg *License:* Public domain *Contributors:* ? *Original artist:* ?

- **File:Australia_satellite_plane.jpg** *Source:* https://upload.wikimedia.org/wikipedia/commons/e/ed/Australia_satellite_plane.jpg *License:* Public domain *Contributors:* The image is from [1], specifically land_shallow_topo_east.tif, which was cropped at 5250x4320+13390+11880. *Original artist:* Reto Stöckl / NASA Goddard Space Flight Center

- **File:Bank_of_Pasadena1895.jpg** *Source:* https://upload.wikimedia.org/wikipedia/commons/7/75/Bank_of_Pasadena1895.jpg *License:* Public domain *Contributors:* Sargent, Shirley 'T.P. Lukens' page 17 *Original artist:* Unknown

- **File:Biodiversity_on_clearcut.jpg** *Source:* https://upload.wikimedia.org/wikipedia/commons/a/ae/Biodiversity_on_clearcut.jpg *License:* Public domain *Contributors:* Source *Original artist:* Office of Surface Mining Reclamation and Enforcement, upload by User:Quadell

- **File:Birdsey_Northrop.jpg** *Source:* https://upload.wikimedia.org/wikipedia/commons/5/5c/Birdsey_Northrop.jpg *License:* Public domain *Contributors:* https://www.flickr.com/photos/internetarchivebookimages/14785197365/ *Original artist:* Internet Archive Book Images

- **File:Blue_Starfish.jpg** *Source:* https://upload.wikimedia.org/wikipedia/commons/8/8e/Blue_Starfish.jpg *License:* CC BY-SA 2.0 *Contributors:* Blue Starfish *Original artist:* Richard Ling

- **File:Broken_trees.jpg** *Source:* https://upload.wikimedia.org/wikipedia/commons/6/66/Broken_trees.jpg *License:* CC BY-SA 3.0 *Contributors:* photo taken by Emilio Bruna *Original artist:* Etwillia

- **File:CO2_increase_rate.png** *Source:* https://upload.wikimedia.org/wikipedia/commons/0/06/CO2_increase_rate.png *License:* CC-BY-SA-3.0 *Contributors:* I created this image with help (numerical data only) from Dr. Pieter Tans (3 May 2008) "Annual CO2 mole fraction increase (ppm)" for 1959-2007," National Oceanic and Atmospheric Administration Earth System Research Laboratory, Global Monitoring Division (additional details; see also K.A. Masarie, P.P. Tans (1995) "Extension and integration of atmospheric carbon dioxide data into a globally consistent measurement record," J. Geopys. Research, vol. 100, 11593-11610.) *Original artist:* New Image Uploader 929 (talk)

- **File:Cecropia.jpg** *Source:* https://upload.wikimedia.org/wikipedia/commons/f/f0/Cecropia.jpg *License:* CC BY-SA 3.0 *Contributors:* photo courtesy of Emilio Bruna *Original artist:* Etwillia

- **File:Commons-logo.svg** *Source:* https://upload.wikimedia.org/wikipedia/en/4/4a/Commons-logo.svg *License:* ? *Contributors:* ? *Original artist:* ?

- **File:Composite_nature_icon.jpg** *Source:* https://upload.wikimedia.org/wikipedia/commons/7/7b/Composite_nature_icon.jpg *License:* CC BY-SA 3.0 *Contributors:* Created as a composite image from 4 other files on Commons: File:Bunch of flowers.svg, File:Greentree.svg, File:Weather-clear.svg, and File:Weather-violent-storm.svg *Original artist:* TFCforever

39.4.3   Content license